H. Planck M. Dauner M. Renardy (Eds.)

Degradation Phenomena on Polymeric Biomaterials

With 85 Figures and 27 Tables

Springer-Verlag
Berlin Heidelberg NewYork
London Paris Tokyo
HongKong Barcelona Budapest

Proceedings of the 4th International Conference on Biomaterials, Denkendorf
September 3 - 5, 1991

Dr. Ing.Heinrich Planck
Dipl.-Ing. Martin Dauner
Dipl. Biol. Monika Renardy

Institut für Textil- und Verfahrenstechnik
Körschtalstraße 26
W - 7306 Denkendorf
FRG

ISBN-13 :978-3-642-77565-9 e-ISBN-13 :978-3-642-77563-5
DOI: 10.1007/978-3-642-77563-5

Softcover reprint of the hardcover 1st edition 1992

Typesetting: Camera ready by authors

02/3020-5 4 3 2 1 0 - Printed on acid -free paper

PREFACE

Traditionally, the International ITV - Conferences on Biomaterials are focussing on problems in Biomedical Engineering, problems, which are still unsolved, of main interest, and which are of interdisciplinary character.

In 1983, the Division of Biomedical Engineering of the Institute of Textile Technology and Chemical Engineering, Denkendorf, started with a conference about the use of polyurethanes in biomedicine.

Three years later, in 1986, progress in development and use of polyurethanes was selected as conference topic. It had to be realized that degradation problems were still dominating the discussions.
The main discussion topics were: What are the causes for the degradation? How can one prevent them? What are the degradation products, and do they affect the human body? How can one simulate the degradation? How can one accelerate the in vitro tests and how can the results predict the in vivo behaviour of the material? How do in vitro tests correlate with animal tests and the behaviour in the human body?

At the third conference in 1989, the speakers focused on the use of textiles in medicine. Again the problem of degradation was discussed intensively and demonstrated by the failure of textile implants, the degradation of aramide polymers or the degradation of resorbable suture materials. The examples make clear, that degradation may be a desired or undesired property of an implant. For the temporary fixation of tissue or bones, a controlled degradation followed by the resorption of the material is wanted. Implants made of nonresorbable materials should show no changes in properties, which means that degradation is undesired and should be avoided; examples are vascular grafts and heart valves. Or, if this is not possible because of the type of application, degradation has to be minimized. An example for this application are hip endoprostheses.

Degradation can be caused by hydrolysis, oxidation or enzymes as well as by mechanical stress. Degradation depends on the type of polymer, the processing, the processing additives needed, and also on the finish of the device. As it is known from the polyurethanes, degradation is strongly related to and sometimes caused by the processing of the material, especially, if it is a thermal processing beyond the melting temperature or if the material is stressed mechanically in the solid state.

The in vivo degradation of implants is not fully understood. That is the reason why we decided to organize the 4th International ITV Conference on Biomaterials as a meeting concentrating on the degradation phenomena of polymeric biomaterials. Following the presentation of test methods for simulation, evaluation and acceleration of biodegradation, the behaviour of nondegradable, degradable, or resorbable materials was discussed in separate sessions. The effects of the processing on biodegradation was the topic of another session. Clinical aspects and experiences completed the presentations.

Each paper was discussed intensively. The discussion remarks are added to each paper. As usual, a round table summarized the results of the conference.

We would like to thank all participants and speakers for the excellent presentations and discussions during the meeting.

Denkendorf, 17. February 1992

The editors

CONTENTS

TESTING OF PUR - BIOMATERIALS FOR ENZYMATICAUY INDUCED DEGRADATION: AN IN VITRO STUDY

B. Glasmacher-Seiler, J. Jonas, G. Schädlich, H. Reul, G. Rau

Helmholtz-Institute for Biomedical Engineering
Pauwelsstraße 30, D-5100 Aachen

Abstract: The enzymatically induced degradation of a group of medical grade polyurethanes (PUR) for cardiovascular applications such as heart valve prostheses and blood pumps has been investigated. Mechanical stability of these polyether-, polyester- and polyurethaneureas has been tested by incubation of microtensile specimens according to ASTM 1708-79 into enzymatic solutions and bovine blood plasma for up to 4 months. The degree of degradation is assessed by measuring tensile properties, water absorption and molecular weight. Changes in surface quality are determined by SEM and contact angle measurements. The results so far reveal degradative changes depending on PUR-type and type of enzyme but to a degree not impairing the implant function within the chosen test period and test conditions.

INTRODUCTION

When we talk of testing of materials for biomedical applications which means investigating the biocompatiblity of these materials, we should always keep in mind that we have to perform screening tests that means testing of a couple of materials in parallel. In addition we should select test methods and test conditions and parameters in the view of the desired application of the materials. Only in this configuration, testing of biocompatibility is meaningful.

Within this study materials are tested which are used to make heart valve prostheses or blood contacting surfaces of blood pumps. For these applications biomedical polyurethanes are investigated. The test parameters and conditions are chosen in an appropriate manner for example with regard to the time period, blood pumps are in contact with the human body. The tests are performed at physiological temperatures and in physiological solutions. Another important parameter adapted to the valve or pump application is the type of specimen fabrication: The samples are solution cast in a similar manner that pumps are made. In addition, the size of the test specimens is comparable. The surface to volume ratio is certainly of importance. Therefore, the test specimens are in the same thickness range of 0.2 mm as pump membranes or valve leaflets.

Three types of polyurethanes are under investigation: polyether-, and polyesterurethanes, and polyetherurethaneureas depending on the components of the polymer (type of polyol, diiso-

H. Planck M. Dauner M. Renardy (Eds.)
Degradation Phenomena on Polymeric Biomaterials

cyanat, and chain extender). The polyurethanes are solved in a wide range of solvents. As the results reveal that the kind of solvent has less influence on the biocompatibility properties that are investigated than the type of polyurethane, this study will concentrate on the latter aspect.

The mechanical and surface properties of biomaterials can be impaired by the action of biological components. This can result in a loss of functionality of the implant and a loss of biocompatibility. This process that leads to a gradual reduction of material properties and qualities in the physiological environment is called biodegradation. The biological environment may act by several processes on the degradation of implants. One possible process is hydrolysis. If water molecules are absorped, covalent bondings in the macromolecule may be splitted resulting in a loss of mechanical stability, molecular weight or a decrease in the degree of cross linkings. The ester- and ether-bonds as well as the urethane- and urea-bonds in the polyurethane molecule can be hydrolysed. The presence of water on its own will not lead to a significant hydrolysis, but if biological catalysts are present, biodegradation may be accelerated. Hydrolytic enzymes may act as catalysts in this context. By their action, the activation energy is decreased thus facilitating hydrolytic splitting of macromolecules. In addition, analogues have been found between urethane bonds and peptide-bonds which are in the natural substrates of the enzymes, namely in polypeptides or proteins. /1, 4 - 11/.

MATERIALS AND METHODS

The test materials include polyesterurethanes from Akzo/Enka, all based on the same polyurethane but solved in different solvents (1017 (DMF), 1025/1 (DMA), 1025/2 (THF/Dioxan), 1076 (Chloroform), 1175 H (DMF)), the polyetherurethane Pellethane^R 2363-80 AE (DMF, THF, DMA) and polyetherurethane ureas (BPS-215 M^R (DMA) from Thoratec Inc.; Pampul-3 + 1% AMEO^R (DMA) from BDF). The following abbreviations are used for the test materials: BPS, Pa, Pe, 1017,1025/1, 1025/2, 1076, 1175 H.

The following enzymes are studied: Papain, trypsin, α-chymotrypsin, α-amylase and urease. They may act on implants because some of these proteolytic enzymes are extracellularily excreated by cells near the implant following implantation because of foreign body reactions of the organism in the wound healing period. Thus, they may be present in higher amounts near the implant surface. The enzymes are additionally selected with regard to their substrate patterns that means, that some of the chemical bonds in the polyurethane molecule are similar to that of the natural substrate.

The enzyme solutions are prepared with comparable enzyme activities of 5,000 U/l. This is impaired by different definitions and measurements of enzyme activity of the various enzymes that are studied. The solutions are all buffered at enzyme pH-optimum in Sörensen sodium phosphate buffer. The solutions are changed weekly. During this period a loss of activity of 10% to 60% has to be taken into account. Papain exhibits the highest stability.

PUR-specimens are incubated in these solutions for up to four months at 37°C. A group of samples is exposed to mechanical loading following incubation at a strain rate of 50% at 15 Hz for 24 hours (-> 1.3 Mio. cycles).

Static tensile tests are performed according to ASTM 1708-79, DIN 500 14-23/50-2, and DIN 53504 and tensile properties before and after incubation are compared. Weight gain or loss, changes in molecular weight and surface quality of the specimens are determined. Contact angle measurements are performed with a goniometer applying the advancing angle-method with different diagnostic solutions (water, glycerol, formamide, thiodiglycol, ethylene glycol, and hexadecane) and evaluated according to Zisman. The critical surface tensions and the water contact angles of incubated and native specimens are compared.

RESULTS

The stress/strain curves of the tested polyurethanes on the whole exhibit a similar tendency towards changes in the mechanical behaviour. The stress/strain curves after 8 weeks and 12 weeks reveal a decrease in elongation at break and modulus, whereas the tensile strength remains more or less unchanged /2/. A comparison of the effects after 8 and 12 weeks of incubation with regard to the applied enzyme reveals that the enzymes trypsin and α-chymotrypsin exhibit an accelerated action on the biomaterials /2/. Figures 1 and 2 show the values for tensile strength and elongation at break for a polyether-, a polyester- and urea polyurethane without treatment, after 12 weeks incubation in papain, after 1.3 million cycles deformation and cyclic deformation following 12 weeks papain incubation. The mechanical properties are not generally reduced by the fatigue testing.

The weight measurements result in weight gains and weight losses. Figure 3 compares the average weight changes after 12 weeks incubation in the various enzyme solutions. Nearly all polyurethanes exhibit average weight losses after 12 weeks incubation in α-amylase (-0.62%) and phosphate buffer (-0.72%) and they gain in weight after incubation in α-chymotrypsin

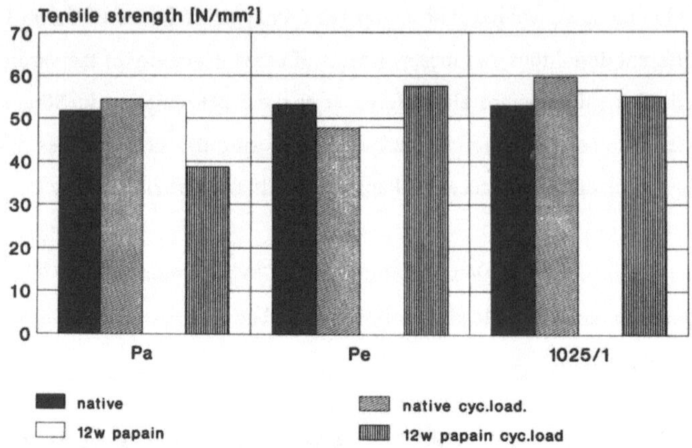

Fig. 1: Influence of 12 weeks papain incubation and mechanical loading on the tensile strength of Pampul[R], Pellethane[R], and PUR 1025/1[R]

(+0.43%). 50% of the polyurethane-specimens lost weight in trypsin by -0.6%, the mean weight gain ist +0.5%. Following incubation in urease 40% of the samples gained weight (+0.2%), 60% lost weight (-0.74%). In papain, 36% had an increase in weight of +0.3% and 64% a decrease of -0.36%. Looking at the averaged results for the individual polyurethane (figure 4) reveals that 80% of the BPS-specimens lost weight by -1.0%, 70% of the Pampul-samples by -0.64%. Only 40% of the Pellethane-samples lost weight by -0.33% and 45% of the polyester-urethanes by -0.28%. In order to find explanations for this behaviour, - from literature data an increase in weight due to water absorption is expected -, water-uptake measurements according to DIN 53495 specifications are performed. They show water uptake saturation in the range of

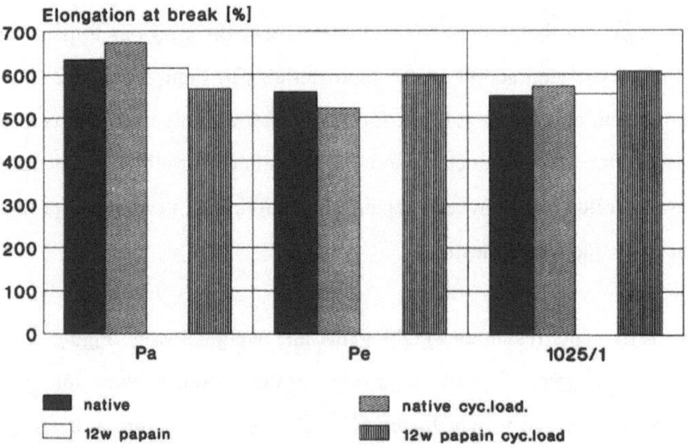

Fig. 2: Influence of 12 weeks papain incubation and mechanical loading on elongation at break of Pampul[R], Pellethane[R], and PUR 1025/1[R]

0.3 to 0.4 wt% after 2 days of incubation in destilled water at room temperature for BPS, Pampul, Pellethane, 1025/1, and 1175H. As these results are comparable to literature data, it has to be considered that the enzymes act differently on the polyurethanes. The polyurethanes are affected by the enzyme incubation depending on the type of polyurethane: The urea-type lost weight whereas the increase in weight of the polyester- and polyetherurethanes was in the range of the values of the water-uptake measurements.

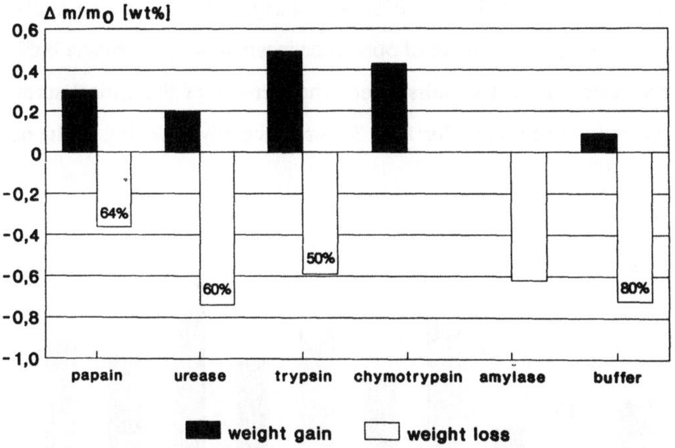

Fig 3: Influence of enzyme incubation on the average PUR-weight change ($\Delta m/m_0$) (the numbers in the bars indicate the percentage of samples exhibiting gain losses)

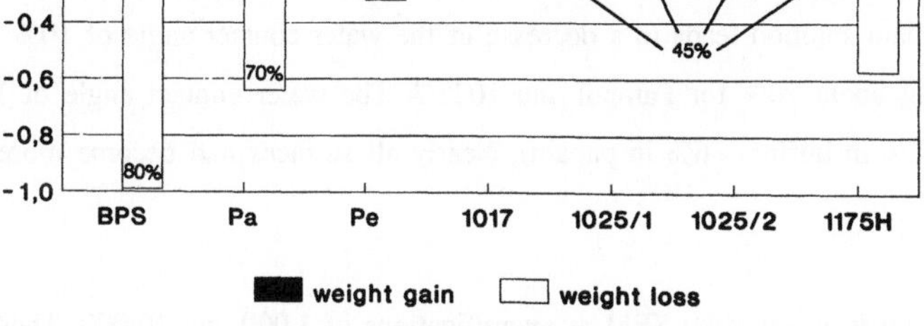

Fig. 4: Average weight changes of various polyurethanes (the numbers in the bars indicate the percentage of samples exhibiting gain losses)

This result is confirmed by molecular weight determinations, performed after four months of incubation at 37°C in papain, trypsin, α-chymoptrypsin, and α-amylase. The results so far show that the molecular weight distribution of the polyetherurethane as well as of the polyester-urethanes are not markedly changed by the enzymes. But there are changes for the urea types. Pampul is affected by trypsin, α-amylase and papain.

The results of the contact angle measurements are shown in figures 5 and 6. The critical surface tension of the air-dried surface of BPS is unchanged, 1025/2 is also rarely affected (decrease in papain of -9%). An increase of 56% after incubation in phosphate buffer and of 37% in papain is evaluated for the Pampul-surface. The increase of the critical surface tension for Pellethane is 5% respectively 27%, for 1017 37% respectively 43% (fig. 5). Incubation in pure

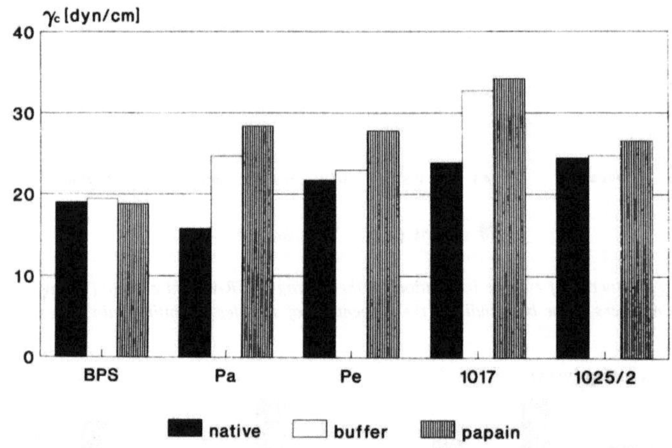

Fig. 5: Critical surface tension γ_c of the air-dried surfaces of various polyurethanes without treatment, and after incubation in phosphate buffer and papain solution

buffer or papain solution leads to a decrease of the water contact angle of 10% for BPS and Pellethane, of about 20% for Pampul and 1025/2. The water contact angle of 1017 is least affected (+ 5% in buffer, - 6% in papain). Nearly all surfaces had become more hydrophilic (fig. 6).

Surface quality is assessed by SEM at magnifications of 1,000 and 10,000. There are micro-bubbles (ϕ 2 μm) and small pores (ϕ 0.5 μm) which are due to the sample casting process, but no morphological changes due to an enzymatic attack.

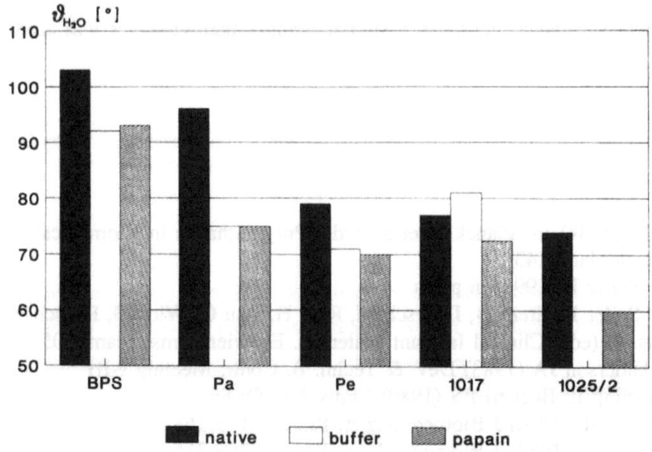

Fig. 6: Water contact angle θ_{H_2O} of the air-dried surfaces of various polyurethanes without treatment, and after incubation in phosphate buffer and papain solution

CONCLUSIONS

The test results show that enzymatically induced biodegradation is a possible mechanism that may act on the implant in vivo. The tested polyurethanes exhibit sufficient biostability during the chosen test period of up to four months as detectable by the selected test methods. Nevertheless, it has to be considered that the in vivo conditions cannot fully be simulated by such in vitro testing methods. As the applied enzyme activities are nearly ten times higher than expected in the in vivo application, the results should be on the "secure side", and the simulation of the in vivo conditions should be justifiable. The degree of degradation is not enhanced by the papain concentration. The effects are the same for 1,000 and 5,000 U/l which has been shown elsewhere /3/. With respect to the comparability of the test sets, all incubations are however performed with an average enzyme activity of 5,000 U/l. A comparison of the influence of bovine blood plasma and papain solution during incubation from 4 to 12 weeks for three different polyurethanes (Pellethane, Pampul, and 1025/1) reveals in addition, that elongation at break and tensile strength do not change significantly; the average change is ± 5%. Only for Pellethane, elongation at break decreases by 7% in plasma and both values decrease by 10% in papain /3/.

None of the tested polyurethanes failed or showed massive changes, ie. reduction in mechanical stabiblity or surface quality. Nevertheless, the results give rise to the assumption that enzymes

varying criteria to determine whether or not a material is biodegradable, the results of different studies are often not comparable. In this paper, biodegradation will be defined as chemical modification of the polymer by chemicals produced by the host. Even if the extent of attack is not great enough to adversely affect mechanical properties, chemical modification may produce leachable materials or alter the interaction of the polymer with proteins and cells.

An additional factor contributing to the differences seen by researchers may lie in the PEUs and not in the techniques used. Small differences in the polymers, undetectable by many conventional chemical analysis methods, may alter the biostability of the polymers. We have investigated the commercial medical grade PEU Biomer™ in order to determine differences between two lots and ascertain the effect these differences have on oxidative and enzymatic degradation of the polymer. Biomer™ has been extensively studied and often used as a "standard" for PEU studies. Biomer™ is a poly(ether urethane urea) and is thought to be manufactured from poly(tetramethylene oxide) (PTMO), methylene-bis phenyl-diisocyanate, and ethylene diamine[8]. Evidence of cyclohexane diamine, diethylamine, BHT, 4,4' -butylidene-bis-(t-butyl-m-cresol), and poly(diisopropylaminoethyl methacrylate co decyl-methacrylate) have also been found in one lot of Biomer™ [9]. Richards et.al. [10] have observed lot to lot differences in the pyrolysis mass spectra of Biomer™. Some lots they investigated contained large fractions of the poly (diisopropylaminoethyl methacrylate) while others contained none. The influence of the these difference, however, on the biostability of the PEUs has not been determined.

We used four methods for measuring differences between the two lots of Biomer™ before and after chemical modification: high pressure gel permeation chromatography (HPGPC), fourier transform infrared spectroscopy (FTIR), x-ray photoelectron spectroscopy (XPS), and static secondary ion mass spectroscopy (SIMS). XPS and static SIMS have proven to be valuable and complementary techniques for studying polymer surfaces[11].

EXPERIMENTAL

Materials

Two lots of solution grade Biomer™, lot BSP067 (received in 1982) and lot BSUA001 (received in 1987), were studied. For the purposes of this article, these lots will be referred to as BSP and BSUA, respectively. Both lots were received from Ethicon Corporation, New Brunswick, NJ, as 30 wt% solutions in n,n-dimethyl acetamide (DMAC). The samples were dried soon after they were received and stored dry at ambient conditions until analysis. Samples from each lot were characterized by using HPGPC, FTIR, XPS, and SIMS.

Gel Permeation Chromatography

HPGPC analysis was performed to determine the molecular weight distribution of the polymers before and after treatment with hydrogen peroxide. A 10 μl injection of each sample was introduced into a DMAC mobile phase pumped at 1.0 ml/min. Two Alltech C-18 derivatized Nucleosil™ columns, one containing 80Å pore size packing and the other containing 1000 Å pore size packing, were used. Monodisperse polystyrene standards gave a linear calibration curve over a molecular weight range from $>10^6$ to 500. A Waters 410 refractive index detector was interfaced to a computer for data collection and calculation of molecular weight distribution. Weight and number average molecular weights (Mw and Mn) were calculated using software developed by Dr. Steven Pollack that used the methods outlined by Yau et al [12]. Each sample was injected five times to permit statistical comparison. A student's t- test at the 95% level was used to determine significance of changes in Mw.

Infrared Adsorption Spectroscopy

Bulk chemistry of the two Biomer™ lots was studied by using transmission FTIR. FTIR spectra were obtained by using a Biorad 60A spectrometer with an MCT detector. Samples of both Biomer™ lots were solution cast onto NaCl crystals from 20 mg/ml (1,1,1,3,3,3)-hexafluoro-2-propanol (HFIP) solution. 20 μl of solution were applied to the crystals and allowed to dry for a minimum of 2 hours at ambient conditions. Spectra were collected for 250 scans at 2 wavenumber resolution.

X-ray Photoelectron Spectroscopy

XPS data for the hydrogen peroxide and deionized water-treated samples were obtained on a Surface Science Instruments Model SSX-100 XPS system. A 5 eV flood gun was used to offset charge accumulation on the samples. Elemental compositions were determined on the basis of peak areas from the C1s, N1s, and O1s orbitals by using spectra collected at a pass energy of 150 eV. At this pass energy, the transmission function of the SSX-100 was assumed to be constant [13]. Signal intensities have been scaled by $KE^{0.7}$ to account for the variation in the sampling depth with the kinetic energy of the electrons [13]. Peak areas were calculated by using a linear background correction. Photo-ionization cross sections were taken from Scofield's calculations[14]. No angular asymmetry correction was made. Both the absolute and normalized signal intensities were calculated with the standard SSX-100 software.

The binding environments for C, N, and O, were determined by using spectra collected at 25 eV pass energy with a 1000 μm x-ray spot size. Binding energies were referenced by setting the C1s hydrocarbon peak to 285 eV. The C1s, O1s, and N1s peak envelopes were fit with Gaussian peaks using the standard SSX-100 software.

Static Secondary Ion Mass Spectrometry

The static SIMS experiments were done on a SubMonlayer SIMS instrument on a Surface Science Instruments Model SSX-100 vacuum chamber. A 4-keV Xe^+ primary beam of ~ 1 nA/cm^2 current density was rastered over a 5mm^2 area of the sample. Total accumulated ion dose per sample was kept below 10^{13} ions/cm^2 during data acquisition to ensure that static SIMS conditions prevailed. A quadrupole mass spectrometer with a mass range of 1-800 was used to detect the secondary ions. Charge neutralization was done by flooding the sample with 700 eV electrons.

Oxidative Treatments

The two Biomer™ lots were exposed to 30% hydrogen peroxide for 24 hours to study oxidative degradation of the samples. Hydrogen peroxide was chosen because it is produced by macrophages during the inflammatory response. Degradation of the two lots by enzymes has also been investigated [15]. The activity of the hydrogen peroxide was measured by titration with $KMnO_4$. Concentrated hydrogen peroxide was used to accelerate degradation so that susceptibility to attack during the inflammatory response could be modeled in a short-term *in vitro* experiment. The conditions were not intended to mimic the *in vivo* environment. Deionized water was used as a control.

Samples from each lot of Biomer™ were dissolved at 50 mg/ml in 1,1,1,3,3,3-hexafluoro-2-propanol (HFIP). Samples for the degradation experiments were prepared by evaporating 40 µl of solution onto clean 9 mm diameter glass cover slips. The cast films weighed between 2-3 mg and were 15 - 20 µm thick.

Samples from both lots were exposed to either hydrogen peroxide or distilled water for 24 hours at 37°C. Following treatment, the samples were dip rinsed in deionized water and then dried overnight in a vacuum desiccator. Treatments were done in quadruplicate. Three of the samples were examined with XPS for surface changes and were then dissolved at 1.0 mg/ml in DMAC and filtered through 0.5 µm Teflon Millipore filters for HPGPC analysis. The fourth sample was used for static SIMS analysis. Because of the destructive nature of the technique, the samples analysed with SIMS were not used for HPGPC analysis.

RESULTS

Chemical Analysis

We have published FTIR, XPS, and SIMS spectra and HPGPC chromatograms for the two Biomer™ lots[16]. Transmission FTIR spectra do not reveal significant differences between the lots. HPGPC did reveal differences in the molecular weight distributions of the two lots. The Mw for lot BSP is 1.5 times the Mw for lot BSUA.

XPS: Table 1 lists XPS elemental compositions for the two lots as a function of electron take-off angle. Lot BSP shows significantly more O and less N than lot BSUA at every take-off angle. Figure 1 shows the C1s peaks for the two Biomer™ lots acquired at an 80° take-off angle. Under these conditions the XPS sampling depth is 15-20Å into the surface. The C1s envelope of lot BSP can be resolved into three peaks, a hydrocarbon peak at 285.0 eV, an ether peak at 286.4 eV and a urea/urethane peak at 289.2 eV [17]. The data indicate that the surface of lot BSP is dominated by the poly(tetramethylene oxide) soft segment with small amounts of the urea and urethane. The O1s and N1s spectra [16] corroborate this observation.

Table 1
Surface Composition from XPS

Lot	angle	%C	%O	%N
BSP067	10°	79.5	18.0	2.4
	55°	80.0	18.2	1.8
	68°	80.3	18.3	1.4
	80°	80.4	18.6	1.0
BSUA001	10°	81.2	15.5	3.3
	55°	81.5	15.0	3.5
	68°	83.0	13.1	3.9
	80°	84.5	10.9	4.5

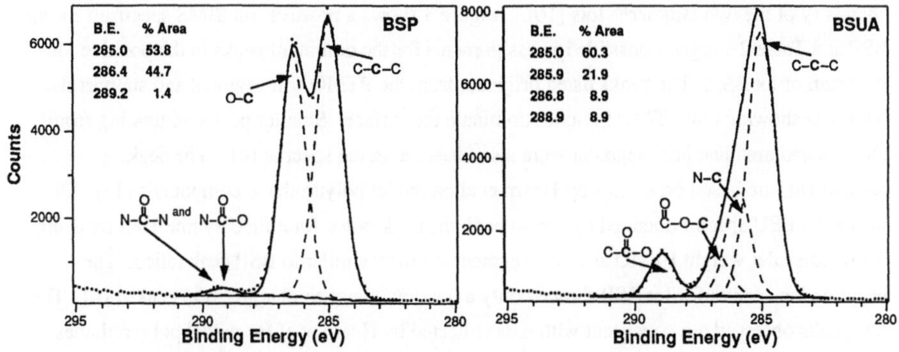

Figure 1. The XPS C1s peak for BSP can be resolved into three component peaks: hydrocarbon (285 eV), ether (286.4 eV) and amide (289.2 eV). The C1s for BSUA can be resolved into four component peaks: hydrocarbon (285 eV), amine (285.9 eV), and esters (286.8 eV and 288.9 eV).

In contrast, the C1s envelope for BSUA can be resolved into four peaks. The binding energies of the peaks suggest hydrocarbon (285.0 eV), amine (285.9 eV), and ester functionalities (286.8 eV and 288.9 eV). The O1s peak shows strong contributions from ester species and the binding energy of the N1s peak indicates that amines rather than ureas or urethanes constitute 85.9% of the surface nitrogen [16].

Depth profiles calculated for angle dependent XPS data are shown in figure 2. The two lots show opposite trends. Lot BSP shows polyether enrichment in the outermost 10 Å. Lot BSUA shows complete disappearance of the ether oxygen at the surface and enrichment of nitrogen, hydrocarbon, and ester groups at the surface. The composition 40 Å below the surface is similar for both polymers.

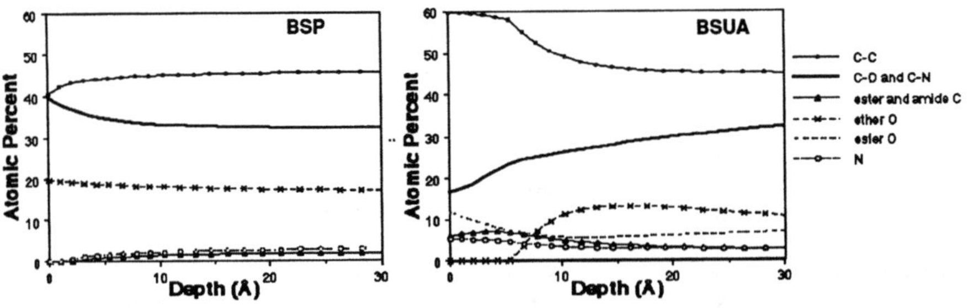

Figure 2. Concentration depth profiles calculated from angle dependent XPS data show the two lots have opposite trends in the surface region. Polyether functionalities are enriched in the upper 5Å-10Å of lot BSP but are completely absent in the upper 5Å - 10Å of lot BSUA.

<u>SIMS:</u> Positive and negative ion SIMS spectra also revealed differences in the surface chemistry of the two Biomer™ lots [16]. Figure 3 shows a positive ion SIMS spectrum for lot BSP and Table 2 suggests possible ion assignments for the dominant peaks in the positive ion spectrum of lot BSP. The peaks listed originate from the PTMO soft segment and support the XPS data showing that PTMO moieties dominate the surface. Smaller peaks originating from the urea and urethane hard segment were also observed in the spectra[16]. The peak assignments are based on studies by Hearn et al on model polyurethane compounds [11]. Of the model PEU spectra obtained by Hearn et al, the peak ratios for a PEU synthesized by using 2000 molecular weight PTMO as a soft segment are most similar to BSP peak ratios. The negative ion spectra for lot BSP showed only a few low mass-to-charge (m/z) peaks [16]. The few peaks observed are consistent with data collected by Hearn et al for model polyurethanes

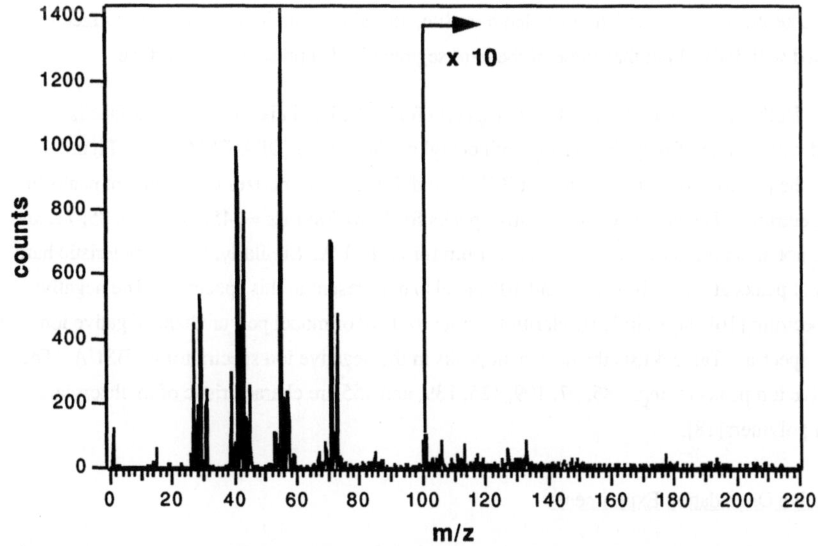

Figure 3. The positive ion SIMS for lot BSP contains peaks characteristic of PTMO, MDI, and an antioxidant.

Table 2
Possible Structures of Positive Secondary Ions from Poly (tetramethylene glycol) Soft Segment

mass	structure
145	$CH_3CH_2CH_2CH_2OCH_2CH_2CH_2CH_2{=}\overset{+}{O}H$
127	$CH_3CH_2CH_2CH_2\overset{+}{O}{=}CHCH_2CH{=}CH_2$
101	$CH_3CH_2CH_2CH_2\overset{+}{O}{=}CHCH_3$ and $CH_3CH_2CH_2O{=}CHCH_2CH_3$
87	$CH_3CH_2CH_2CH_2\overset{+}{O}{=}CH_2$ and $CH_3CH_2CH_2\overset{+}{O}{=}CHCH_3$
85	$C_5H_9O^+$
73	$C_4H_8OH^+$
71	$C_4H_7O^+$
55	$C_3H_3O^+$ and $C_4H_7^+$

[11]. Like the angle-dependent XPS depth profile, the SIMS data indicate that the surface is enriched with PTMO but that some of the hard segment is also present at the surface.

Both the positive and negative ion spectra for lot BSUA indicate that the surface is coated with a layer of poly(diisopropylaminoethyl methacrylate) (DPA-EMA) [16]. Figure 4 shows the positive ion spectrum for lot BSUA, and Table 3 summarizes the dominant peaks in the spectrum. The characteristic positive peaks for PTMO at m/z = 145, 127, 100, 85, 71, and 55 are not apparent in the positive ion spectrum for lot BSUA. Similarly, the characteristic hard segment peaks at m/z = 165, 132, and 106 are also not present in this spectrum. The negative ion spectrum [16] shows little similarity to either lot BSP or model polyurethane negative ion SIMS spectra. Table 4 lists the dominant peaks in the negative ion spectra for lot BSUA. The negative ion peaks at m/z= 85, 97, 109, 125, 139, and 155 are characteristic of methacrylate-based polymers[18].

<u>Oxidative Degradation Experiments</u>

Statistically significant changes in Mw were observed for both lots when treated with hydrogen peroxide. No change in Mw was observed for the deionized water controls. Figure 5 shows a typical chromatogram for samples of lot BSP treated with water and hydrogen peroxide. The number of chains with greater than 10^6 molecular weight in lot BSP increased after the peroxide treatment. This indicates that crosslinking is the dominant chemical reaction of lot BSP with hydrogen peroxide. The Mw of lot BSP increased 9.1% as a result of peroxide treatment which was significant at the 95% level. In contrast, the Mw for a typical sample of BSUA decreased 20.6% when exposed to hydrogen peroxide, indicating that chain cleavage is the dominant reaction. Figure 6 shows that the peak of the molecular weight distribution is lowered by 3×10^6 and a shoulder has formed in the mw greater than 10^6 region. These results were confirmed qualitatively in each of three replicate samples. The extent of degradation varied about 20% between the replicates as a result of non-uniformities in the cast films.

Evidence of chemical modification of the surfaces treated with hydrogen peroxide could be seen in the XPS results. From these XPS studies we found that samples treated with hydrogen peroxide had surfaces richer in oxygen than those treated with deionized water. Changes were also seen in the XPS N1s peak of samples treated with hydrogen peroxide. Figure 7 shows the changes in the N1s peak for lots BSUA. The N1s peak, for samples of BSUA treated with water, consists of two components: an amine peak (399.0 eV), and a urea/urethane peak (400.0 eV). The amine contribution is from the DPA-EMA additive. For samples of lot BSUA treated with peroxide, the N1s peak contains contributions from four nitrogen species: amine (399.0 eV), urea and urethane (400.0 eV), protonated amine (401.3), and nitroso (403.0). The ratio of the N1s urea/urethane peak area to the C1S peak area did not change with peroxide treatment but the ratio of the N1s amine peak area to the C1s peak area was reduced by 58%. The

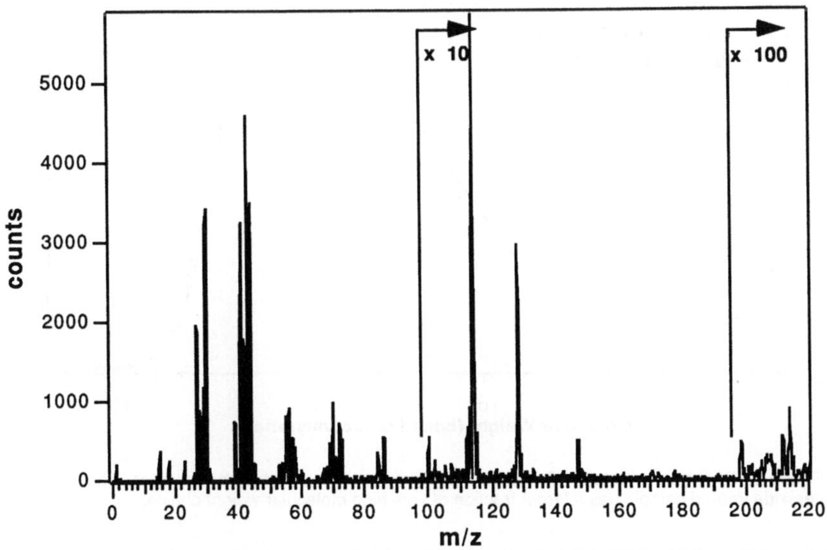

Figure 4. The positive ion SIMS for lot BSUA contains peaks for the DPA-EMA additive. Characteristic ions for PTMO and MDI are absent.

Table 3
Possible Structure of Positive Secondary Ions from DPA-EMA additive

mass	structure
214	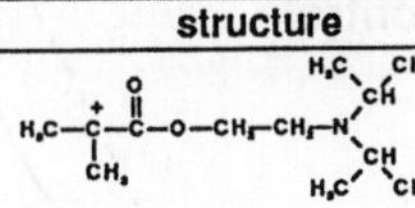
212	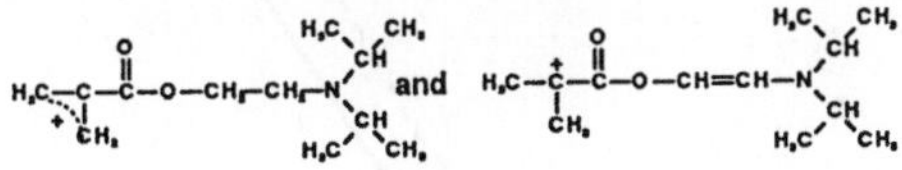
128	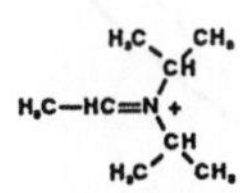
114	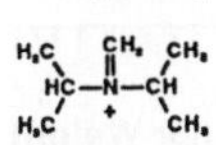
100	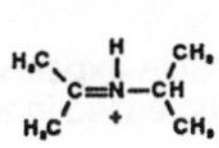
86	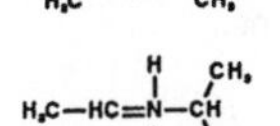
72	$(CH_3CH_2)_2-N^+$
44	$CH_3CH_2\overset{+}{N}H$ or $H_3C-\overset{+}{N}-CH_3$

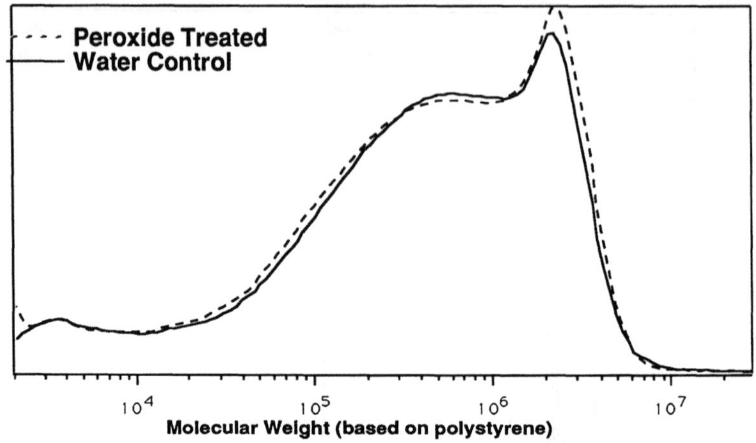

Figure 5. GPC traces for lot BSP exposed to hydrogen peroxide and distilled water. The peroxide treated sample has a larger fraction of very high molecular weight chains.

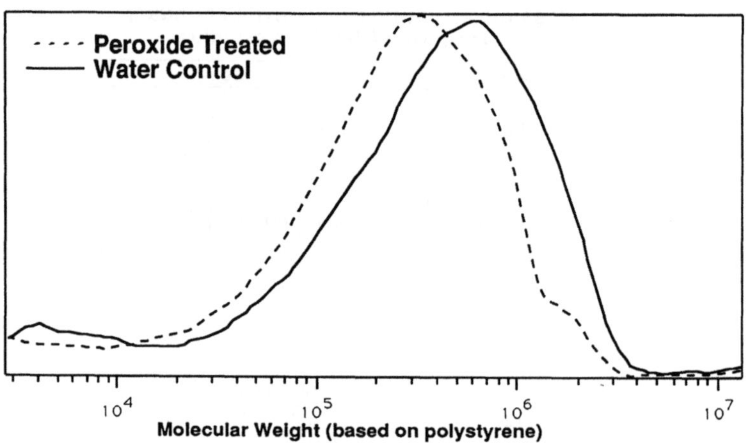

Figure 6. GPC traces for lot BSUA exposed to hydrogen peroxide and distilled water. The peroxide treated sample has decreased in average molecular weight and shows a high oss of very high molecular weight fractions..

protonated amine and nitroso species were formed from the DPA-EMA additive and not from the urea or urethane linkages in the polymer. Tertiary amines, like DPA-EMA, react with oxygen free radicals to form nitroso radicals.

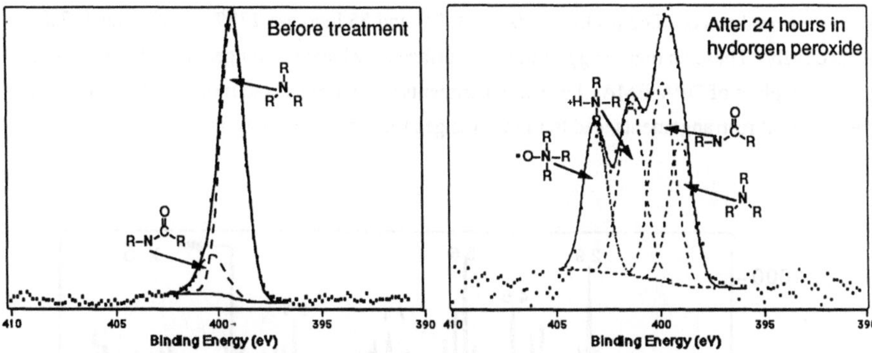

Figure 7. The XPS N1s spectra for lot BSUA changes after peroxide treatment. Before treatment 2 peaks: an tertiaryamine peak (399 eV) and a urea/urethane peak (400 eV) are present. After exposure to hydrogen peroxide two additional peaks are present in the N1s spectra: a quaternary amine peak (401.3) and a nitroso peak (403).

The static SIMS results also show evidence of chemical modification at the surface of both samples that were exposed to hydrogen peroxide. Figure 8 shows the difference between a SIMS spectrum of lot BSP before peroxide treatment and a SIMS spectrum of BSP after peroxide treatment. The spectra were normalized to the hydrocarbon peak at m/z = 27.

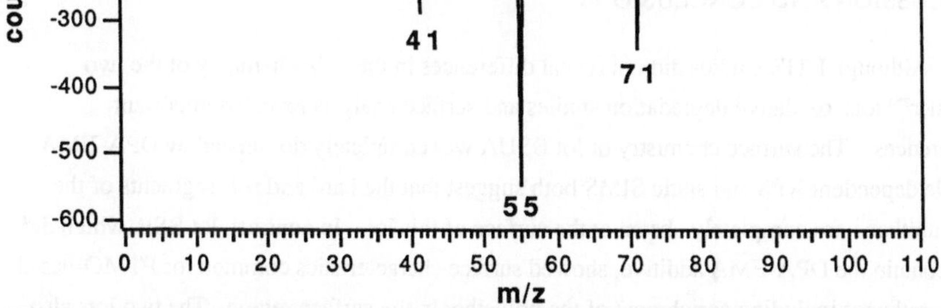

Figure 8. The difference between SIMS for lot BSP before and after exposure to hydrogen peroxide shows several peaks have decreased relative to the m/z= 27 peak. The intensity of the peaks at m/z = 71 and 55 have decreased 45% and 37% respectively.

Figure 9 shows the difference between SIMS spectra of lot BSUA before and after peroxide treatment. The peaks at m/z = 114 and 128 decrease 82% and 79%, respectively, after peroxide treatment. All of the amine containing peaks decrease even though the surface nitrogen content, based on XPS measurements, remains constant. Significant increases are as seen in several peaks. The peaks at m/z = 91, 71, and 55 increase 278%, 128%, and 157% respectively. These changes suggest that the surface may have rearranged, bringing PEU to the surface in place of DPA-EMA. Further experiments are in progress to more precisely determine the chemical rearrangements lead to in the changes seen in SIMS analysis.

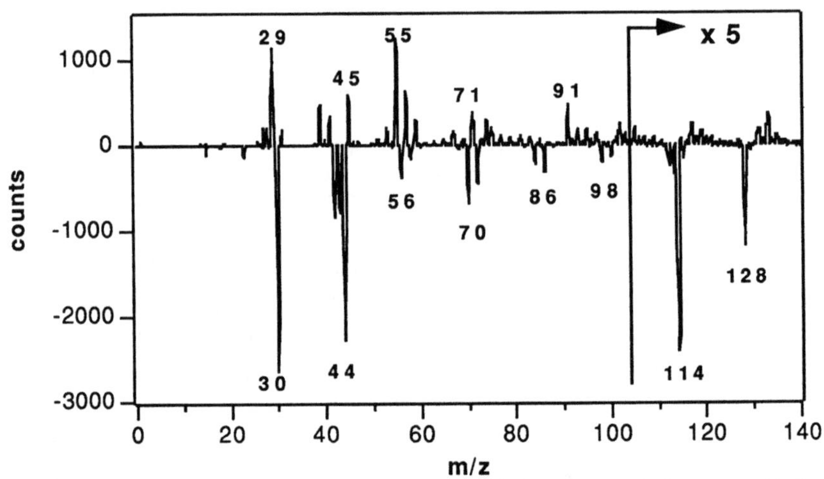

Figure 9. The difference between SIMS for lot BSUA before and after exposure to hydrogen peroxide shows both peaks which have increased and decreased in intensity relative to the peak at m/z=27. The DPA-EMA peaks at m/z=114 and 128 decrease. Peaks at 91, 71 and 55 increase and may indicate presence of the PEU at the surface.

DISCUSSION AND CONCLUSIONS

Although FTIR studies did not reveal differences in the bulk chemistry of the two Biomer™ lots, oxidative degradation studies and surface analysis revealed important differences. The surface chemistry of lot BSUA was completely dominated by DPA-EMA. Angle dependent XPS and static SIMS both suggest that the hard and soft segments of the polyurethane are completely absent at the surface of this lot. In contrast, lot BSP, which did not contain the DPA-EMA additive, showed surface characteristics common for PTMO-based polyurethanes including enrichment of the polyether in the surface region. The two lots also reacted differently when exposed to concentrated hydrogen peroxide. The Mw of lot BSP increased 9.1% but the Mw of lot BSUA decreased 20.6%. It is unclear whether the two lots reacted differently with hydrogen peroxide because of the DPA-EMA additive or if the oxidative degradation studies have revealed an independent difference between the two Biomer™ lots. Preparative-scale HPGPC is currently being used to prepare samples of Biomer™ lot BSUA which are free of the DPA-EMA additive. These samples will be used to investigate the possibility of additional differences between the two lots.

Static SIMS and XPS analysis revealed chemical modification of both polymer surfaces. Nitroso groups and protonated amines were formed from the DPA-EMA additive on the surface of lot BSUA. SIMS analysis also suggested rearrangements of the surface that reduced the ion yield from the amine side chain and possibly increased the amount of PEU at the surface. XPS showed incorporation of oxygen into the surface of lot BSP and SIMS suggested that cleavage of the PTMO chains was occurring.

For both lots, the chemical reactions occurring at the surface are different from the reactions observed for the bulk. Biomer™ will absorb between 10% and 17% hydrogen peroxide by weight [15], so opportunity for the hydrogen peroxide to react directly with the PEU far below the surface exists. Lot BSUA showed significant cleavage of the PEU chains in the bulk, but at the surface, only attack of the DPA-EMA was observed. Lot BSP was significantly crosslinked in the bulk but the surface showed evidence of chain cleavage. These difference are at least in part due to the differences between bulk and surface chemistry of the polyurethanes but may also be caused by differences in reaction rates and mass transfer rates. The results indicate that hydrogen peroxide can undergo at least two different reactions with polyurethanes, one resulting in chain cleavage and one resulting in crosslinking. Under some conditions a one reaction dominates but under others the other may dominate.

This study demonstrates that differences between PEUs, too small to be detected by FTIR, can affect the rates of degradation reactions. The existence of competing reactions also suggests the possibility that different reaction mechanisms may exist in the acute and chronic inflammatory responses or that different hosts or different implantation sites may cause different reactions to dominate. *In vivo,* the PEU would be attacked simultaneously by enzymes and oxidants. In addition, extraction of components, absorption of lipids and other biomolecules, and adsorption of protein would occur. Crosslinking of the polymer may result in brittleness, and extraction may cause surface flaws. Swelling of the polymer may then lead to stress crazing. Small cracks forming in the polymer could then serve as passages giving enzymes access to the bulk of the polymer. Each of these factors may shift the balance between competing oxidative reactions with the PEU. The ultimate effect *in vivo* is difficult to predict. A better understanding of the kinetics of these reactions and the factors that affect the kinetics are needed in order to understand how PEUS behave *in vivo*.

REFERENCES

1. Hunter SK, Gregonis BE, Coleman DL, Andrade JD, and Kessler T, (1982) Trans Am. Soc. Artif. Intern. Organs, 28,473 - 477

2. Marchant RE, Anderson JM, Phua K, and Hiltner A (1984) J. Biomed. Mater. Res., 18, 309-315

3. Zhao Q, Marchant RE, Anderson JM, and Hiltner A, Polymer 28, 2040-2046, 1987.

4. Marchant RE, Zhao Q, Anderson JM, and Hiltner A (1987) Polymer 28, 2032-2039,

5. Hayashi K, Matsuda T, Takano H, and Umezu M (1984) J. Biomed. Mater. Res., 18, 939-951, 1984.

6. Takahara A, Coury AJ, Hergenrother RW, and Cooper SL, (1991) J. Biomed. Mater. Res., 25, 341-356

7. Ratner BD, Gladhill KW, and Horbett TA (1988) J. Biomed. Mater. Res. 22, 509-527, 1988

8. Richards JM, McClennen WH, Meuzelaar HLC, Gregonis DE, Reichart M, and Helle M (1985) Macromolecules, 18, 496-500

9. Belisle J, Maier SK, and Tucker JA (1990) J. of Biomed. Mat. Res., 24, 1585-1598

10. Richards JA, McClennen WH, and Meuzelaar HLC (1990)J. of Appl. Polym. Sci. 40, 1-12

11. Hearn MJ, Ratner BD, and Briggs D (1988) Macromolecules, 21, 2950-2959

12. Yau WW, Kirkland JJ, and Bly DD (1979) Modern Size-Exclusion Liquid Chromatography, John Wiley and Sons, 1979, Chapter 10

13. Applications Note, Surface Science Instruments (1987)

14. Scofield JH (1976) J. Elect. Spectrosc. Relat. Phenom. 8, 129-137

15. Tyler BJ and Ratner BD (1991) J. Biomed Mater. Res. submitted

16. Tyler BJ, Ratner BD, Castner DG, and Briggs D (1991) J. Biomed. Mater. Res., in press.

17. Castner DGand Ratner BD (1988)J. Vac. Sci. Technol., 6 (3), 964-965,1988.

18. Hearn MJ and Briggs D (1988) Surface and Interface Analysis, Vol. 11, 198-213 (198

SELECTED DISCUSSION REMARKS:

Brandwood: Would you like to speculate, what the changes are that are taking place in Biomer under storage?

Tyler: The primary thing that we think to see under storage is the increase in the molecular weight. There seems to be some crosslinking that goes on.
The stuff is much more stable when you dry it, instead of keeping it in solution. So we took samples, we dried them and kept them refrigerated in the dark. The difference that we have seen was the additive. The additive is a polymer and you can't extract it out with any solvent we have in our laboratory. We are still trying to track down exactly why they degrade so differently but I really don't think it is age-related.

IN-VITRO DEGRADATION TEST FOR SCREENING OF BIOMATERIALS

C. Welling, H. Schwengler, B. Strahl

Beiersdorf AG
Unnastraße 48, D-2000 Hamburg 20

SUMMARY

Specimens made of three different biodegradable materials (BDF 1, BDF 2 and BDF 3) were exposed in-vitro to an enzyme-cocktail under mechanical stress for 1, 2 and 3 weeks. Degradation was determined by measuring dry weight loss, mechanical properties (compression test) and SEM surface analysis. BDF 2 samples were found to degrade in such a way, that considerable weight loss was measured, fissures at material surface were observed and the specimens became softer after three weeks incubation. BDF 1 and BDF 3 samples were found to lose weight but there was just a slight degradation observed by SEM analysis at the material surface. After three weeks incubation the samples became stiffer, so that pressure needed for 20% deformation increased.

INTRODUCTION

Biodegradation is defined as the gradual breakdown of a material mediated by a specific biological activity (1). In-vitro studies about the interaction of polymers with different enzyme solutions suggest, that there is a possibility of enzyme induced degradation (2-3-4-5). The mechanical stress upon an implant is known to produce in some biopolymers fissures or cracks at the material surface. These fissures or cracks serve as sites for enzymatic attack and further degradation. However, there are still some open questions about the exact course of biodegradation. The search for an appropriate material to be used as an implant in the ileum region of the human intestine lead to the conclusion, that besides chemical charateristics, the biological behaviour of the material had to be tested. For this purpose we have developed a test system for in-vitro degradation, which approximately simulates the enzymatic and physico-mechanical conditions of the intestinal region. The

H. Planck M. Dauner M. Renardy (Eds.)
Degradation Phenomena on Polymeric Biomaterials

material had to be flexible enough to withstand the peristaltic movements of the intestine and on the other hand, it had to keep its form and stability over a period of at least two to three weeks.

EXPERIMENTAL

Material: Three different materials, which were defined as degradable by the manufacturers, were used under Code Names BDF 1, BDF 2 and BDF 3 in our in-vitro degradation test.

Preparation of samples: Injection moulded specimens were prepared from the a.m. materials. The specimens had cilindrical form with a length of 16 mm and a wall thickness of 0,3; 0,4 and 0,5 mm. After moulding, the specimens were sterilized by γ irradiation (28 KGray) and used for test purpose.

METHODS

Enzymes: The experiments were carried out with an enzyme cocktail, composed by the enzymes trypsin and chymotrypsin. These enzymes were dissolved in phosphate buffer at a pH of 7,6. Enzyme concentration was as following:

Trypsin	:	1,0 mg
Chymotrypsin	:	1,0 mg
Tris-buffer	:	ph 7,6 ad 100 ml

The solution was stabel for at least 24h. The enzyme activity was measured at an elapsed experimental time t=0 h and t=24 h at 37°C

Trypsin	=	2.500U/mg
Chymotrypsin	=	5U/mg

Every 24 hours, the solution was removed and fresh solution was filled in.

Mechanical test procedure: The test device illustrated in Fig. 1 consists of several silicon tubes which were placed between pads. The pads were subdivided into chambers. The chambers were filled subsequently with air, so that the peristaltic movement in the ileum was simulated. The samples which had to be tested were carefully inserted in these silicone tubes. Subsequently the enzyme solution was pumped through the cylindrical silicone tubes. The test was run over a period of time of 1,2 and 3 weeks.

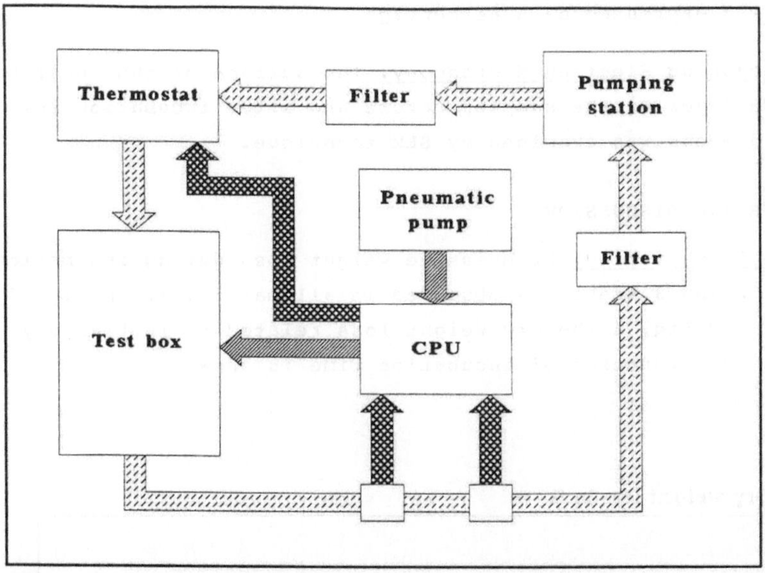

Fig 1. Mechanical test device

Determination of degradation: The grade of degradation was measured by determining weight loss, mechanical properties and by examining the surface of the test samples. The results were compared to the results which were obtained by measuring the same parameters on the control specimens which were kept at room temperature.

Weight loss: Test samples and control samples were dried to constant weight, before and after 1, 2 and 3 weeks incubation in the test system. Weight loss was expressed as percental dry weight loss related to the initial dry weight of the samples.

Mechanical properties: compression test

All tests concerning determination of mechanical properties of the specimens were carried out on a Frank testingmachine. In this machine, the specimens were fixed and the force needed for a compression of 20% of the test specimens determined. Changes in mechanical properties were defined by determining the increase or decrease of force that was needed for a compression of 20% of the test specimen in camparison to the control. The ratio was expressed as a percentage.

SEM: Scanning Electron Microscopy: The surface of the inner and outside layer of the samples before and after incubation for 1, 2 and 3 weeks was examined by SEM technique.

RESULTS AND DISCUSSION

Weight loss studies: Progressive weight loss during incubation for 1, 2 and 3 weeks was observed in all materials. In fig. 2, fig. 3 and fig. 4 the dry weight loss related to initial dry weight as a function of incubation time is shown.

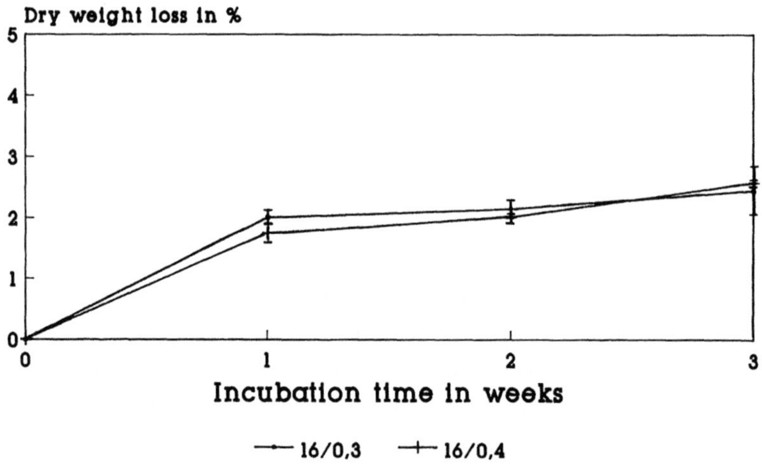

Fig 2. Dry weight loss (%) BDF 1 (16/0,3; 16/0,4)

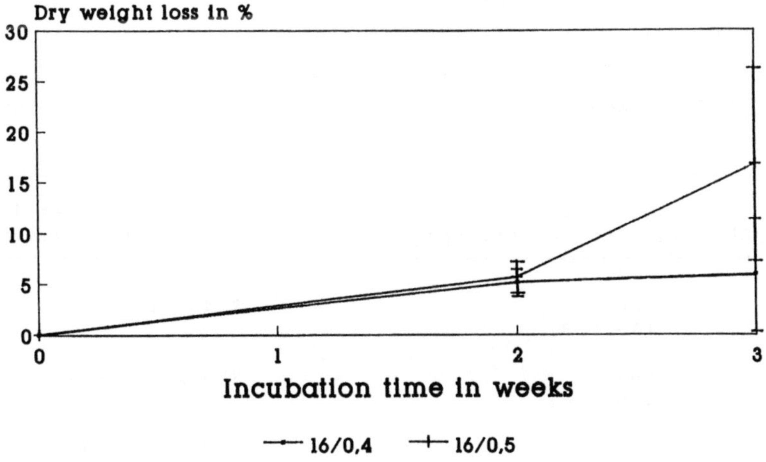

Fig 3. Dry weight loss (%) BDF 2 (16/0,4; 16/0,5)

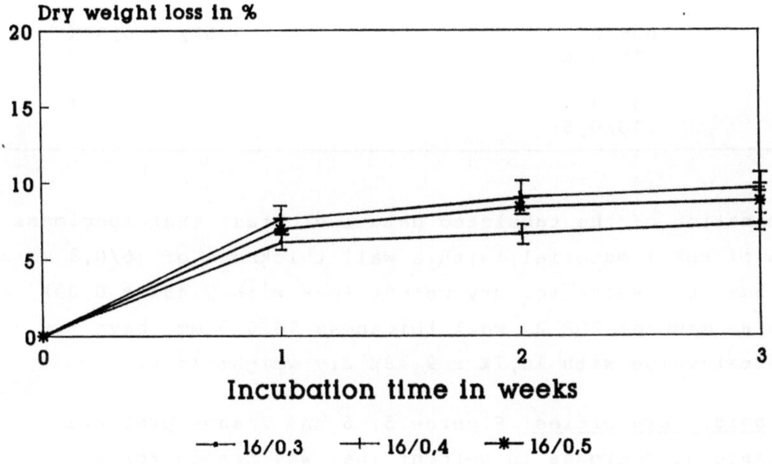

Fig 4. Dry weight loss (%) BDF 3 (16/0,3; 16/04; 16/0,5)

As can be seen the dry weight loss of the samples made of BDF 1 and BDF 3 was almost independent from wall thickness. In order to allow a comparision between the Specimen produced from BDF

1, BDF 2 and BDF 3 the experimental results concerning dry weight loss are listed in table 1.

Tab. 1: Dry Weight loss (%) after three weeks incubation (n=16)

Material (length / wall thickness. [mm])	Dry weight loss (%) (mean, standard deviation)
BDF 1 (16/0,3)	2,45% ± 0,09
BDF 1 (16/0,4)	2,58% ± 0,41%
BDF 2 (16/0,4)	5,9 % ± 5,7%
BDF 2 (16/0,5)	16,7% ± 9,48%
BDF 3 (16/0,3)	7,32% ± 0,48%
BDF 3 (16/0,4)	9,62% ± 0,37%
BDF 3 (16/0,5)	8,80% ± 1,62%

Examination of the tabulated data indicates, that specimens made of BDF 1 material, with a wall thickness of 16/0,3 mm show the smallest value for dry weight loss with 2,45% ± 0,09%, and samples made of BDF 2, wall thickness 16/0,5 mm, have the greatest value with 16,7% ± 9,48% dry weight loss.

Mechanical properties: Figures 5, 6 and 7 show pressure increase or decrease in percent that was needed for a compression of 20% of the test specimens in comparison to controls. As can be seen for samples made of material BDF 1 and BDF 3, with progressing incubation time, the samples became stiffer, and there is a continuous pressure increase needed for 20% deformation. In BDF 1 materials, the wall thickness of the specimens has almost no influence on the mechanical properties. Samples made of BDF 2 become softer during incubation time, as as the pressure decrease which is needed for 20% deformation of the specimen decreases.

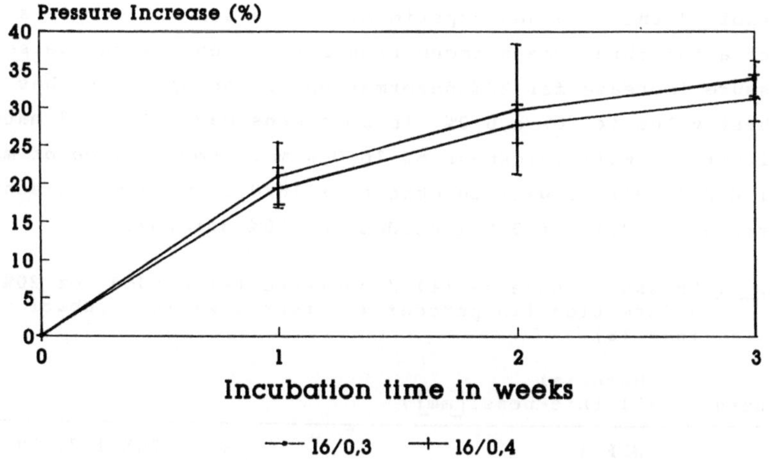

Fig 5. Pressure Increase (20% Deformation) BDF 1 (16/0,3; 16/0,4)

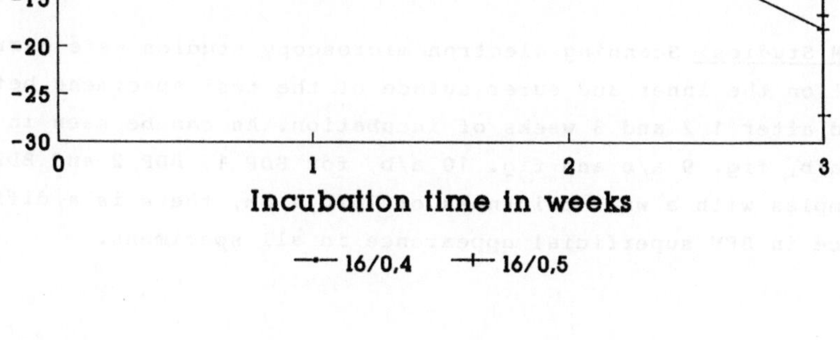

Fig 6. Pressure Decrease (20% Deformation) BDF 2 (16/0,4; 16/0,5)

In table 2 the obtained experimental data for all compression tests after three weeks incubation are listed. As can be seen, pressure increase for 20% deformation of the specimens has its highest value (41,0% ± 6,0%) in specimens made of BDF 3 material with a wall thickness of 16,0,4 mm. Samples made of material BDF 2, with a wall thickness of 16/0,5 mm show a pressure decrease of 18,12% ± 9,02% needed for 20% deformation.

Tab. 2: Pressure increase (+) / decrease (-) needed for 20% deformation (in percent %) after 3 weeks incubation (n = 16)

Material (length / wall thickness. [mm])	
BDF 1 (16/0,3)	+ 33,78% ± 2,49%
BDF 1 (16/0,4)	+ 31,13% ± 3,26%
BDF 2 (16/0,4)	- 8,53% ± 7,96%
BDF 2 (16/0,5)	- 18,12% ± 9,02%
BDF 3 (16/0,3)	+ 7,32% ± 0,48%
BDF 3 (16/0,4)	+ 9,62% ± 0,37%
BDF 3 (16/0,5)	+ 8,80% ± 1,62%

SEM Studies: Scanning electron microscopy studies were carried out on the inner and outer suface of the test specimens before and after 1,2 and 3 weeks of incubation. As can be seen in fig. 8 a/b, fig. 9 a/b and fig. 10 a/b, for BDF 1, BDF 2 and BDF 3 samples with a wall thickness of 16/0,4 mm, there is a difference in SEM superficial appearence in all specimens.

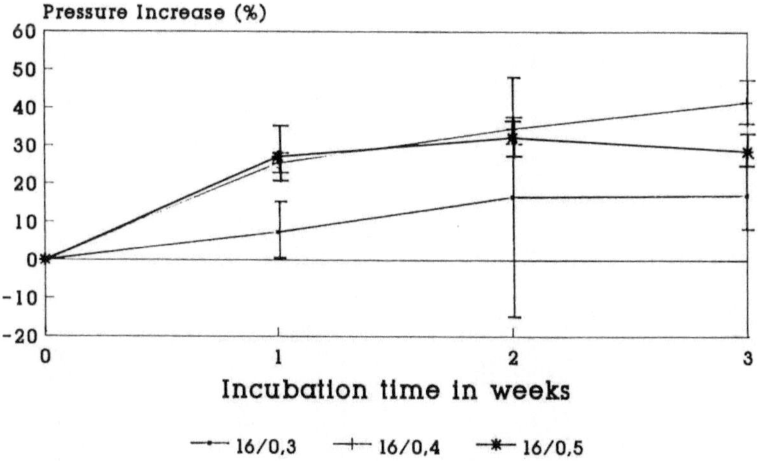

Fig 7. Pressure Increase (20% Deformation) BDF 3 (16/0,3; 16/0,4; 16/0,5)

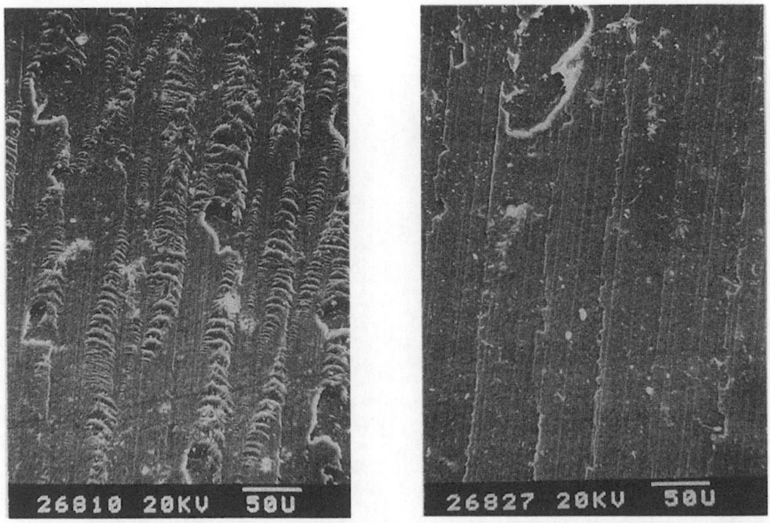

Fig 8. SEM Surface analysis before (8a) and after incubation for three weeks (8b) BDF 1 (16/0,4)

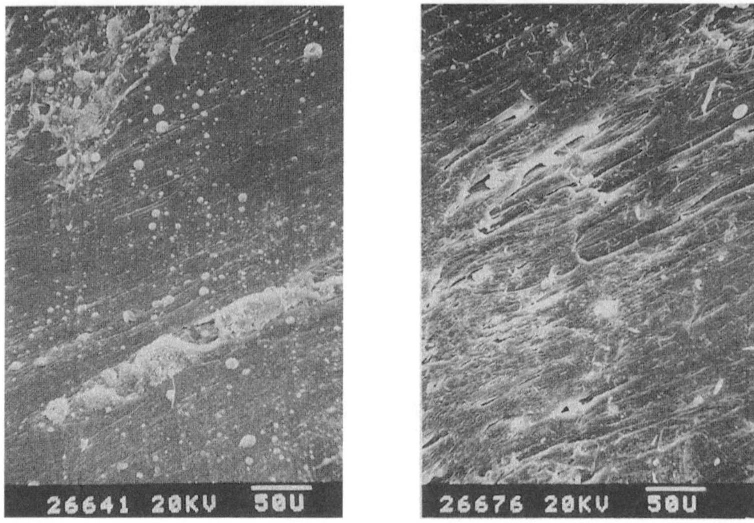

Fig 9. SEM Surface analysis before (9a) and after incubation for three weeks (9b) BDF 2 (16/0,4)

Fig 10. SEM Surface analysis before (10a) and after incubation for three weeks (10b) BDF 3 (16/0,4)

After 3 weeks, material BDF 1 and BDF 3 show slight degradation in form of missing traces of the injection moulding procedure. Samples made of BDF 2 material however, show fissures on the surface and of course, the traces of injection moulding procedure are also missing.

GENERAL DISCUSSION AND CONCLUSIONS

In order to establish a comparison between all materials that were used for the experimental runs in the in-vitro test system, the results after three weeks incubation, that means, dry weight loss (%), mechanical properties and SEM surface analysis for the specimens with a wall thickness of 16/0,4 mm are listened in table 3.

<u>Table 3:</u> Results after three weeks incubation (dry weight loss %), mechanical properties and SEM studies. (n=16)

Material (length / wall thickness)	dry weight loss (%)	Percent pressure increase (+) decrease (-) for 20% deformation	SEM
BDF 1 (16/04)	2,58% ± 0,41%	+ 31,13% ± 3,26%	slight degradation
BDF 2 (16/04)	5,9 % ± 5,7 %	- 8,53% ± 7,96%	fissures cracks
BDF 3 (16/04)	9,62% ± 0,37%	+ 9,62% ± 0,37%	slight degradation

As described by Holland et al. (6) injected moulding specimens of diffenent biomaterials are very stabel to biodegradation. On the other hand, Miller et al. (7) described, that weight loss is just one way of degradation. Decrease or change of mechanical properties of a material can occur without having a simultaneous weight loss. A comparison of the experimental data listened in table 3 lead to the conclusion, that for our research purposes, a degradable material which looses weight continuously, gets softer, and at the same time mantains its initial form is optimal. We selected the material BDF 2 for further research and development.

REFERENCES

1 Williams D.F. (1987) Definitions in Biomaterials (D.F. Williams, ed.) Elsevier 4:68

2 Williams D.F. (1977) Enzyme-Polymer Interactions, Journal of Bioengeneering I:279-294

3 Smith R, Oliver C and Williams D.F. (1987) The enzymatic degradation of polymers in-vitro, Journal of Biomedical Materials Research 21:991-1003

4 Phua S.K., Castillo E, Anderson J.M. and Hiltner A (1987) Biodegradation of a polyurethane in vitro, Journal of Biomedical Materials Research 21:231-246

5 Schakenraad J.M., Nievwenhuis P., and Molenaar I., Helder J., Dijkstra P.J., and Feijen J (1989) In-vivo and in-vitro degradation of glycine / DL-lactid acid copolymers, Journal of Biomedical Materials Research 23:1271 - 1288

6 Holland S.J., Jolly A.M., Jasin M., and Tighe B.J (1987) Polymers for biodegradable medical devices, Biomaterials 8 pp:289-295

7 Miller M.D. and Williams D.F (1987) On the biodegradation of poly-ß-hydroxybuyrate (PHB) homopolymer and poly-ß-hydroxybuty-rate-hydroxyvalerate copolymers 8:129-137

THE BIODEGRADATION OF NONDEGRADABLE POLYMERS

K.B. Stokes, P.W. Urbanski

Medtronic, Inc.
7000 Central Ave.
Minneapolis, MN, 55432, USA

INTRODUCTION

The biostability of materials used for long term implantable devices has been taken for granted far too often in the past. It is commonly stated that certain materials are inert, nondegradable or implantable, that is, biostable. Is there really any such thing as a nondegradable, inherently biostable material? Silicone rubber, for example, is a polymer that is generally accepted as biostable. None-the-less, it can absorb lipids and stress crack when used as a heart valve poppet. Silicone rubber finger joints can crack and break. Silicone rubber insulated pacemaker leads can fail because of wear, cold flow and/or calcification. The lives of many people depend upon the resistance of their polyester vascular prostheses to degradation. None-the-less, we have seen examples of fibrous lumps of tissue removed from pacemaker patients containing fragments of the dacron pouch used to keep their pulse generators from migrating. Each of these (and many other) materials has been tested on the bench and has been proven to be stable, resistant to oxidation and other degradation mechanisms. The materials have been implanted in animals for long periods of time and have been proven to be "nondegradable". In fact, these materials have been used very successfully in many chronically implanted device models. Then why the occasional failure of a device made from a biostable polymer? One answer may be the failure or inability to account for device-material-manufacturing process-tissue interactions as well as individual differences (biologic variance).

The reliability of implanted devices for extended periods of time begins with the materials used in the device, but certainly does not end there. The goal of this paper is to review the state-of-the-art in polymer biostability and biodegradation mechanisms as related to the devices they are used in. Given the scope of the subject, the review must necessarily be limited to a few of the more common polymers, devices and examples of material-device failure. Since our experience has been primarily with pacemaker leads and polyurethanes, their biostability and failure mechanisms will be explored in more detail.

H. Planck M. Dauner M. Renardy (Eds.)
Degradation Phenomena on Polymeric Biomaterials

AROMATIC POLYESTER

Polyethylene terephthalate (PET) woven or knitted fabrics are widely use in large caliber vascular grafts, sewing rings for cardiac valves, AV shunts, tissue fixation, etc. In general, the experience with PET fabrics appears to be good, but not entirely trouble-free.[1] In one example, a PET femoro-popliteal vascular graft developed aneurysm after 84 months implant, and was removed. Physico-chemical analysis of the material showed that dilitation had taken place subsequent to hydrolytic degradation of the fibers.[2] Berger, et al did a larger study of 493 PET vascular prostheses of 4 different types with up to 15.3 years implant time.[3] No defects were detected prior to 3 years implant. Deterioration occurred in 15 out of 493 grafts as seen by thinning and breakage of filaments causing holes. Dilatation occurred, but did not always precede filament breakage, which was associated with crimp ridges. The conclusions of this study were that manufacturing variations reduced fiber resistance to mechanical fatigue. The degradative effects of tissue fluids and enzymes were considered to be possible contributing effects.

Canine studies have shown PET degradation rates comparable to those seen in humans. Zaikof found a mean absorption time of 30 $\pm$ 7 years for woven patches with 50% loss of fiber strength in 10 $\pm$ 2 years when implanted in the canine subcutis.[4] In infected sites, hoerver, degradation was accelerated exponentially to produce complete loss of physical properties within a few months. The pH at infected sites was found to be as low as 4.8.

In our own experience we have had several incidences where abandoned cardiac pacemaker leads were accused of provoking a proliferative fibrous response at their subcutaneously buried proximal ends. Examination of the excised tissues, however, revealed fragments of fibers from a PET fabric (Parsonnet pouch) which had been used to provide fixation for the pulse generator. Of course, since the system was abandoned, the pulse generator had been removed some time previously. The untoward biologic response was identified by the pathologist as a response to the disintegrating fabric, not the abandoned lead.

POLYAMIDES

Polyamides, especially nylons are very interesting for implantable devices because of their excellent toughness. However, the in vivo hydrolytic degradation of polyamides such as nylon 6 (polycaproamide) or nylon 6,6 (polyhexamethylene adipate) is well known. For example, Kopecek, et al found that nylon 6,6 lost 25% of its tensile strength after 89 days canine implant and 83% after 726 days.[5] In another study, McClellan et al examined rabbit corneoscleral sections containing nylon sutures after 23 months implant.[6] Only two of nine nylon sutures remained. Hayasaka, et al found that the location of the sutures in the eye was important to biostability.[7] Scanning electron microscopic examination of nylon suture used for corneoscleral wound closure was smooth whereas that used to fix the intraocular lens against the iris showed rough, cracked surfaces. Tests in various biologic media including ocular tissue extracts lead to the conclusion that nylon suture thread may be hydrolysed by lysosomal enzymes and should be considered a late-degrading suture.

Nylon 6 was used to coat the multifiliment tail string of an IUD. In a study of these devices that had been implanted 2 years and longer, Hudson, et al found that 67% had cracks in the tail string coating.[8] These cracks allowed access of bacteria to the multifiliment, in turn serving as a conduit for the bacteria to bypass the protective cervical mucus barrier and enter the sterile uterine cavity.

POLYACETAL

Polyacetal has been used in biomedical applications for over 25 years. Today it has generally been replaced in mechanical heart valves by pyrolytic carbon because of concerns about its durability. Teoh and coworkers compared identical St. Vincents Mechanical valves with polyacetal and pyrolytic carbon occluders in accelerated life cycle tests.[9] Analysis of wear in the Bjork-Shiley Polyacetal valve in vivo for more than 17 years was also done. The results showed that the wear rates of the pyrolytic carbon and polyacetal are similar. Thus, it was concluded that the durability of the occluder was more a matter of mechanical design than material selection per se, and that a superior valve could be designed with the polymer.

POLYOLEFINS

POLYETHYLENE:

Polyolefins such as polyethylene are generally considered to be biocompatible and biologically resistant or biostable polymers. Thus, polyethylene was introduced as a cardiac pacing lead insulation material in the 1960's as an alternative to silicone rubber. While silicone rubber had performed well in early pacing leads, it was felt that the mechanical properties of polyethylene (see table 1) would make the devices more resistant to implant damage, while allowing the them to be made smaller in diameter, yet durable for long implant periods.[10,11]

TABLE 1: SOME COMPARATIVE PHYSICAL PROPERTIES OF PEROXIDE CURED SILICONE RUBBER AND POLYETHYLENE USED IN CARDIAC PACING LEADS

PROPERTY	PEROXIDE CURED SILICONE RUBBER[a]	POLYETHELYNE[b]
Min. Tear Strength (N/m) ASTM D 624, Die B	0.43	---
Young's Modulus (Kg/cm^2) ASTM D 638	43	9500-10500
Tensile Strength (Kg/CM^2) ASTM D 412	84	140-290
Elongation at Break (%) ASTM D 412	450	600-650
Coef. Friction in Blood ASTM D 1894	0.75	0.17

a MDX 4-4515 (Dow Corning).[12]
b Typical medium density polymer.[13]

Subsequent clinical experience, however, showed that the biostability of polyethylene in that end use did not meet expectations. Analysis of more than 100 human explants ranging from 7 days to 11 years showed gradual structural changes with time as well as apparent mechanical deterioration.[14,15] Tensile strength increased initially, then began to decrease after about 20 months. Tensile elongation decreased in proportion to implant time. The material lost half its initial elongation after 33.5 months implant and 100% after 77 $\pm$ 5 months. Cracks or "pittings" were found that, in some cases, completely breached the insulation. The cause of these cracks was shown to be wear of the inner surface of the tubing against the metallic (tinsel wire) conductor. Environmental stress crack tests in simulated body fluids for up to 20 months were negative. Infrared analysis of the explanted tubing revealed distinct changes, including the formation of ester, ketone and perester. Based on a comparison of in vitro thermal oxidation rates with the findings on device explants, it was concluded that the polyethylene oxidizes at much faster rates in vivo than at comparable temperatures in vitro.

Evidence that polyethylene is degraded in vivo and that the degradation products are metabolized and excreted was provided by Openheimer.[16] The first traces of C^{14} in the urine of experimental animals was found 6 months after implantation with radiolabled polyethylene.

Ultra high molecular weight polyethylene (UHMWPE) has been used extensively in joint prostheses. Mechanical deterioration of UHMWPE has been observed in total hip systems because of wear and loading.[17] Wear, creep and oxidative degradation are accelerated by the tendency for frictional heating during articulation of the prostheses.[18] It was shown that ceramic articulating systems generate less frictional heating than polished cobalt alloy against UHMWPE. Such frictional heating is expected to be a function of patient age, weight and activity with the younger, heavier and more active person being more likely to accelerate degradation of the polymeric portion of the device. The production of very tiny particulates resulting from wear of the polyethylene surface has also been observed.[19] In a study of 250 explanted hip cups after from 3 weeks to 14 years implantation revealed that the UHMWPE underwent oxidative chain scission as determined by infrared spectroscopy and increased density.[20,21]

It is commonly believed that polymers such as polyethylene are not degradable by enzymatic processes. None-the-less, Wasserbauer, et al demonstrated that under laboratory conditions, polyethylene foils can be oxidized in the microsomal fractions of bacteria and mouse liver with the formation of carbonyl, ester and hydroxyl groups.[22] It was demonstrated that the oxidation occurred in the polymer matrix per se, not only in oligomeric components. The cause of this oxidation was traced to the monooxygenase hydroxylation system. The key agent shown to be responsible (directly or indirectly) was cytochrome P-450, which can be stimulated with doses of phenobarbital and inhibited by the vaccine propionibacterium acnes and its pyridine fraction.[23]

<u>POLYPROPYLENE</u>

Prolene is a commonly used suture material. In general, it is considered a "permanent" implant. None-the-less, after $\geq$ 1 year in the human eye, Prolene and Perlon both show severe biodegradation in one study[24]. Apple, et al reported good performance of proline loops in intraocular lenses in general with relatively incomplete follow-up 5 to 10 years postimplant[25]. Even so, one device failure after 4 years was found to have cracking of the deeply embedded prolene loop which was interpreted to be the result of stress cracking or oxidation. In contrast, in vitro ultraviolet radiation tests and the analysis of human explants after up to 4.8 years led Mowbray, et al to predict that polypropylene fiber in the anterior chamber of the eye should last from 53 years in Miami to 75 years in Seattle[26].

POLYSILOXANE (SILICONE RUBBER)

Silicone rubber has a relatively long and successful history as an implanted material. None-the-less, mechanical failures of various devices have occurred. For example, in one series of 71 silicone rubber wrist arthroplasties performed over a 7 year period, 20% were found to be fractured and an additional 5% had to be revised for pain and deformity[27]. Of the fractured devices, Twelve of 14 required revision due to wrist pain, instability and deformity. Eleven of these were replaced with another silicone rubber implant and 2 of these refractured. While the specific cause of the failures was not identified, contributory parameters appeared to be excessive wrist motion, overuse and inadequate surgical technique.

Silicone rubber heart valve poppets in one design were found to crack crack by dehiscence.[28] At first it was thought that the part absorbed lipids resulting in cracking. It was later determined that the lipid absorption and cracking occurred as a result of the cyclic stresses imposed upon the occluder as it struck the restraining cage struts.

The use of silicone rubber as an interpositional material in the temporomandibular joint has been questioned.[29] Migration of the material in adjacent tissues led to reoccurrence of symptoms 12 to 18 months after meniscal replacement with silicone rubber in 8% of the cases. Excised tissues, a preauricular lymph node and the explanted material were analysed. Silicon produced prominent peaks in energy dispersive X-ray microanalysis in both pseudocapsular and nodal tissues. Silicone debris induced pathologic responses in the surrounding tissues and migrated to the nodes. The pathologic changes included synovitis, dystrophic calcification, fibrocartilaginous metaplasia, hyalinization and scarring.

In our own experience, silicone rubber cardiac pacemaker lead insulation occasionally fails as a result of cold flow or wear against solid objects such as bones or the pulse generator. Wear has also been seen where one lead section rubs against a second lead or against itself. A relatively new insulation failure mechanism is wear or cut-through by the first rib and clavicle after the device has ben implanted through an introducer via a "subclavian stick" procedure[30]. We also find that silicone rubber cardiac pacing leads can calcify and embrittle over long implant times, resulting in insulation fracture. The actual incidence of failure due to mineralization is unknown, but appears to be relatively small.

POLYESTER POLYURETHANES

Polyester polyurethanes have been investigated extensively for use in chronic implants since the 1950's.[31,32] Their use in plastic and reconstructive surgery initially appeared promising with minimal acute inflammation. Thin fibrous capsules formed with good fixation due to ingrowth into pores. However, they degraded and became fragmented within months as a result of hydrolytic decomposition of the adipate ester soft segments. Similarly, polyester polyurethanes were extensively investigated for the artificial heart program. Their unique toughness, flexibility and blood compatibility made them ideal candidates for blood pump diaphragms. Again, however, hydrolytic degradation of the ether soft segment rendered these polymers unsuitable for chronic implant.[33]

While it has been known for at least 3 decades that these polymers are not biostable, their use in polyurethane foam coated silicone mammary prostheses has continued until only just recently. Apparently, the polyester polyurethane foam covering produced a more desirable fibrotic capsule than the smooth walled silicone implants, even though the foam disintegrated. Today, however, there is considerable controversy over the biocompatibility of the degrading polymer. The concern is that hydrolysis of the urethane linkage can release toluene diamine (TDA), a known carcinogen.[34,35,36]. The question remaining to be answered is whether or not TDA can be formed and released in amounts and rates even close to that required to provoke cancer.

POLYETHER POLYURETHANES

The polyether polyurethane elastomers offer a unique set of physical and biologic properties that make them of intense interest to the biomedical device designer.[37] They have a wide range of elastic moduli combined with excellent toughness, high tensile strength and good tensile elongation. Their toughness results in excellent resistance to tear and wear. Their biocompatibility, and especially their blood compatibility are excellent. They also tend to have low friction when wet which (when combined with their other properties) makes them ideal candidates for devices such as cardiac pacemaker leads, artificial heart diaphragms, vascular grafts, indwelling catheters, and a host of other implantable devices.

There are several polyether polyurethanes available for medical use. All have polytetramethylene oxide (PTMO) soft segments. Pellethanes 2363-80A, 55D, 90A and 75D (Dow Chemical) are thermoplastics with diphenyl methane diisocyanate

(MDI)/butane diol (BD) hard segments and isocyanate chain termination. Other Pellethane 2363 resins have hydroxyl chain termination and are labeled, for example, 80AE. The 80A, 55D, etc. notations refer to the Shore hardness of the polymers while the "E" designates "extrusion grade". The differences in hardness (modulus, etc.) are achieved by varying the hard to soft segment ratio. Pellethanes 2363-80A, 90A & 55D (P80A, P90A & P55D) have been widely used as implantable cardiac and neurologic pacemaker lead insulation. The harder Pellethane 2363-75D has been used extensively as pacemaker connecter housings. Tecoflex EG80A & EG60D (Thermedics) and Estane 5714 (Goodyear) are also thermoplastics. Tecoflex is similar to Pellethane except that dicyclohexyl methane diisocyanate replaces the MDI. Estane 5714 is believed to be similar to the Pellethanes. Biomer (Ethicon) and its analogues have been extensively studied for use in a wide range of implantable devices. The polymer has PTMO soft segments and MDI hard segments, chain extended with ethylene diamine to form urea linkages. Because the urea linkages can degrade at temperatures below thermal processing temperatures, Biomer must be applied from solvent. Mitrathane (PolyMedica), another Biomer-like polymer is used in implantable reinforcement fabrics. Still another Biomer analogue, Surethane (Cardiac Control Systems) is used to insulate some implantable cardiac pacemaker leads. Cardiothane-51 (Kontron) is another solvent based polymer of significant interest that has a relatively long history of use in intra-aortic counterpulsation balloons. Cardiothane-51 is believed to be a moisture curing copolymer of PTMO soft segment, MDI/BD hard segment and (silica free) silicone soft segments with terminal triacetoxy functionality.

THE POLYETHER POLYURETHANE INSULATED PACEMAKER LEAD

So far, the only long term implantable experience with polyether polyurethanes has been as pacemaker lead insulation with the vast majority using P80A, some using P55D and P90A.[38] These polymers were selected because of their unique physical properties compared to silicone rubber as shown in table 2. Polyurethane (P80A) insulated cardiac pacing leads were market released in 1980 after significant in vitro testing, canine implants of devices, long term biostability studies of the material, 2 years very successful human clinical study and a 5 year successful human experience with P55D insulated neurologic leads.[39,40,41] A two year in vivo study of P80A and P55D proved their biostability.[42] Literature searches revealed no known failure mechanisms that could be expected in vivo. The conclusion that the materials were in fact biostable was supported by the work of Boretos, demonstrating the biostability of a structurally similar polymer, Biomer.[43,44]

TABLE 2: SOME COMPARATIVE PHYSICAL PROPERTIES OF SOME TYPICAL SILICONE RUBBER AND POLYETHER POLYURETHANE ELASTOMERS

PROPERTY	PEROXIDE CURED SILICONE RUBBER[a]	Pt CURED SILICONE RUBBER[b]	P80A[c]	P55D[d]
Min. Tear Strength (N/m) ASTM D 624, Die B	0.43	0.91	2.7	3.7
Young's Modulus (Kg/cm^2) ASTM D 638	43	22	245	1430
Tensile Strength (Kg/CM^2) ASTM D 412	84	115	420	460
Elongation at Break (%) ASTM D 412	450	680	550	430
Coef. Friction in Blood ASTM D 1894	0.75	---	0.04	0.04

a MDX 4-4515 (Dow Corning).[13]
b Q7-4650 (Dow Corning).[45]
c Pellethane 2363-80A (Dow Chemical, Formerly Upjohn).[46]
d Pellethane 2363-55D (Dow Chemical, Formerly Upjohn).[40]

In spite of state-of-the-art premarket testing, 2 previously unknown failure mechanisms were later discovered clinically. The failures affected some devices significantly, some slightly and others not at all. Thus, it became clear the the problem was not with the polyurethane per se, but with some lead models made from the polyurethane.

ENVIRONMENTAL STRESS CRACKING (ESC)

Visual and Electron Microscopic Observations:

In vivo ESC on explanted devices is characterized by rough walled cracks that rarely show evidence of crazing.[47,48] The cracks appear to be primarily the result of a tearing, rather than a craze driven mechanism.[49] They occur in patterns that can always be related in orientation and depth to some strain vector in the material. For example, if the polymer is stretched over a noncompressible object such as a metal electrode shank, a hoop stress related to a circumfrential strain will result. In this case, deeper stress cracks will be seen predominantly in the axial dimension (perpendicular to the strain) with shallow cracks at 90°. Most insulation failures were found adjacent to tight ligatures applied to the lead body for fixation. The depth of the cracks as well as the time to failure appear to be directly related to the magnitude of the strain (stress), but Byersdorf demonstrated on human explants that the cracking is not time dependent.[50]

Analytical Results:

Stress-strain measurements of material with shallow surface cracks or of material between deep cracks revealed no significant differences compared to uncracked material.[49,51] Specimens with a shallow notch made with a scalpel, however, had significantly reduced tensile strength and elongation. Thus, the cracks are not stress risers. Some analyses have shown reduced molecular weight in cracked areas, others have shown no change. The explanted polymer has an oxidized surface in the cracked areas which has led many to assume that the mechanism is oxidative stress cracking.[52,53] No other evidence of degradation by chemical reactions has been found.

Accelerated Stress Cracking Tests And The Importance of Manufacturing Processes:

Based on observations made on explanted devices and the hypothesis that cracking was related to the presence of residual strain (stress) in the polymer, a test method was developed using strain as the accelerator.[54] Early testing as a function of implant time and residual strain levels demonstrated that the mechanism in P80A was characterized by induction periods and critical strains that varied significantly with manufacturing processes related to thermal history. For example, P80A tubing extruded with a relatively cool melt (but not cool enough to produce melt fracture) into a cool water tank (maximizing heat transfer rate) cracked very soon after subcutaneous implant in rabbits at low strains. Tubing extruded with a hot melt into a hot tank (Minimizing heat transfer rate) took significantly longer to crack at higher strains. Specimens annealed to relieve 100% of the applied strain had no cracking over the 12 week experiment at fixed elongations of up to 500%.[51] Subsequent animal tests of annealed P80A specimens showed no ESC after 2 years implant except for microscopically shallow cracks where tissue pressure applied stress to the specimen. It was also observed that surfaces oxidized with microscopically shallow pitting and etching. Interestingly, however, only 50% of the animals produced microcracking and oxidation. The other 50% of the animals produced no microcracking visible at up to 5000 magnifications and no oxidation detectable by surface infrared spectroscopy. Thus, it was established that the mechanism had the properties of a form of stress cracking, was modifiable by manufacturing processes and exhibited biologic variance.

The effects and interactions of manufacturing processes were evaluated quantitatively in a 10^6 factorial accelerated in vivo ESC test using 576 specimens. This proved that the overwhelming causative factor was stress, and that annealing to 100% stress relaxation was the key process to prevent ESC, with P values in the 10^{-6} region. Numerous additional significant effects and interactions ($P < .05$) were identified. Unfortunately, removing residual stress from the device as manufactured does not mean that there will be no stress in vivo. Leads can be implanted with kinks and bends in them and there are cyclic strains that cannot be avoided. For example, portions of the lead flex with every heartbeat (almost 104,000 times a day at 72 beats per minute or about 38×10^6 times per year). Thus, while manufacturing process improvements have dramatically reduced ESC failure, ESC still occurs, especially as microcracks that have no effect on the performance of the device.

The Stress Cracking Mechanism:

All the polyether polyurethanes available today still have a common PTMO soft segment. Accelerated in vivo ESC test results on various polymers (including all of the polyether polyurethanes mentioned above) varies as a function of ether content (soft segment molecular weight).[54] P80A with a 1000 Dalton soft segment is more succeptble to ESC than P55D with a 650 Dalton soft segment. Tecoflex EG 80A with a 2000 Dalton soft segment is much more succeptable than P80A. The results also appear to correlate (at least qualitatively) with accelerated in vitro tests in 1 M $AgNO_3$ done by Coury, et al which show that oxidation varies as a function of the ether content.[55,56] Thus, those polymers with high ether content are more succeptable to ESC and oxidation appears to be a part of the mechanism.

With one very recent exception, in vivo stress cracking has not been duplicated in vitro (and we are not yet sure of the exception).[57] No ESC has been observed in vitro in blood, serum or plasma, yet it does occur in in vivo in blood and in cage experiments where the medium is initially essentially plasma, then later serum.[58] What is missing in vitro, but present in vivo is cellular inflammation. Zhao, et al proved that ESC requires the adhesion and activation of phagocytic inflammatory cells, primarily macrophages and foreign body giant cells.[59,60] Activation of macrophages is known to produce the lysosomal release of an incredible number of enzymes and oxidants (as shown in table 3) including hydrogen peroxide, hydroxyl radical and super oxide anion.[61]

TABLE 3: MAJOR PRODUCTS KNOWN TO BE RELEASED BY (ALVEOLAR) MACROPHAGES[61]

Cytokines:
- Interleukin-1 alpha and beta
- Interleukin-6
- Tumor necrosis factor
- Alpha and gamma interferon
- Colony-stimulating growth factors
- Transforming growth factor-beta
- Fibroblast growth factor (insulin growth factor I)
- Neutrophil chemotactic factor
- Neutrophil-activation factor
- Enzyme-releasing factor
- Platelet-derived growth factor
- Histamine releasing factor

Enzymes:
- Lysozyme, beta-glucuronidase
- Acid hydrolases
- Angiotensin converting enzyme
- Elastase; Serine and metalloenzyme
- Collagenase
  - Fibroblastlike
  - Type V (gelatinase)
- Plasminogen activator
- Cysteine proteinase (cathepsin L)

Biologically active lipids:
- Cycloxygenase metabolites
  - Thromboxane A_2
  - Prostaglandins E_2, D_2, 6-keto-PGF-1-alpha and F2-alpha
- Lipoxygenase metabolites
  - 5-hydroxyeicosatetraenoic acid
  - Leukotriene B_4, C_4 & D_4
- Platelet activating factor

Oxygen Metabolites:
- Super oxide anion, hydroxyl radical, H_2O_2

Proteins:
- Antiproteases
  - $Alpha_1$-proteinase inhibitor
  - $Alpha_2$-macroglobulin
  - Plasminogen activator inhibitor
- Other inhibitors
  - IL-1 inhibitor
  - Neutrophil migration inhibitor
  - inhibitor of fibroblast growth factor
  - Lipomodulin
- Glycoprotein
  - fibronectin
- Complement components (C_2, C_4)
- Binding protein
  - Transferrin, ferritin
  - Apolipoprotein

Free fatty acids

Antioxidants:
- Glutathione

Coagulation Factors:
- Factors V & VII

1,25 dihydroxyvitamin D3

Many hypotheses have been put forward to explain ESC. While ESC appears to involve oxidation and oxidants are implicated, in vivo ESC has not been duplicated in vitro at strains of up to 400% in oxidizing media including hydrogen peroxide, hypochlorite and unstabilized (constantly generated) super oxide anion & hydroxyl radical. The oxidized surface phenomena seen on explanted leads has been duplicated in vitro using unstabilized super oxide anion (but not the stress cracks) in both our own work and that of Anderson, et al.[62] Enzymatic activity has often been postulated as the cause of ESC. While mild damage to polyether polyurethanes has been demonstrated in vitro in the presence of certain enzymes, ESC has not been demonstrated.[63,64,65] Still another hypothesis has been that absorption and chelation of calcium ion by the ether soft segment causes embrittlement and cracking of the polymer.[66,67] However, in our studies we find that immersion in aqueous calcium salts and compounds does not reproduce ESC and the calcium found on explanted, cracked surfaces washes off with water. In addition, the very rare examples of calcified polyurethane lead surfaces we examined have been neither embrittled nor cracked. Thus, ESC can not be calcification, oxidative or enzymatic stress cracking.

A hypothesis has evolved that ESC might be a two-staged mechanism. This would involve surface oxidation to produce a crack initiating microdefect followed by plasticization to reduce "yield" strength, allowing the microdefect to propagate into a crack by tearing. Deep cracks with relatively smooth wall morphology have been generated by immersion in hypochlorite solution followed by immersion in an intralipid medium.[57] The differences in the appearance of the hypochlorite/intralipid cracks and in vivo ESC could be due to differences in crack propagation rate. The other phenomena related to in vivo oxidation (pitting & etching) were not duplicated in this experiment. Hypochlorite is generated in vivo by neutrophils through the H_2O_2-myleoperoxidase system.[68] Macrophages, however, contain little to no myleoperoxidase, and are believed to be incapable of generating HOCl on their own.[69,70] The cells capable of releasing hypochlorite (neutrophils) do not appear to be involved significantly in the healing response or the chronic fibrous capsule always associated with ESC on pacemaker leads.[71] The "intralipid" solution used is essentially soybean oil. Thus, it is not yet clear how or whether this work relates to in vivo ESC. Studies to determine the effects of in vitro generated activated oxygen species followed by implant or in vitro immersion in plasma or serum on unannealed, strained specimens are in progress. If this verifies the two stage mechanism, then in vivo ESC can be properly renamed oxidative-mechanical stress cracking.

Meanwhile, we have determined that ESC is a very complex mechanism requiring the interaction of a surprisingly large number of effects as shown in table 4. Elimination of any one element of these interactions will prevent ESC.

TABLE 4: KNOWN ELEMENTS (EFFECTS) BELIEVED TO REQUIRE INTERACTION TO PRODUCE IN VIVO ESC

Residual strain (stress) in excess of a critical value (the depth of the cracks and the induction period are directly related to the magnitude of the residual strain (stress)).
aliphatic Ether linkages.
Morphology placing ether linkages on the surface.
Adhesion and activation of phagocytes, primarily macrophages and foreign body giant cells.[59,60]*
Oxidation of the surface ether by lysosomal oxidants, most probably superoxide anion and/or hydroxyl radical.*
Absorption of a plasticizing agent, identy so far unknown.

* An alternative hypothesis is that the cells causing ESC are neutrophils and the oxidant is hypochlorous acid.[57]

Metal Ion Oxidation (MIO):

Cardiac pacing leads are made of metallic conductor coils surrounded by polymeric insulation as shown in figure 1. Unipolar leads (1 electrode in the heart) have one such coil and one insulator. Bipolar (2 electrodes in the heart) polyurethane leads have two coiled metallic conductors in a coaxial configuration with one insulator between two conductors and one on the outside.

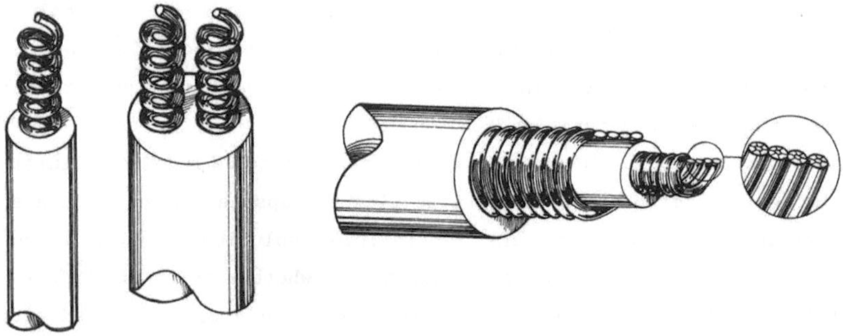

Figure 1: The configuration of three types of pacemaker lead conductors, unipolar (left), bipolar parallel (center) and bipolar coaxial (right).

Today all leads have conductors made of MP35N (Ni, Co, Cr & Mo), Elgiloy (Co, Cr, Ni & Fe) or platinum. The first generation polyurethane insulated leads, however, used a silver/MP35N composite called DBS (named after its manufacturing process, drawn-braze-strand) as shown in figure 2.

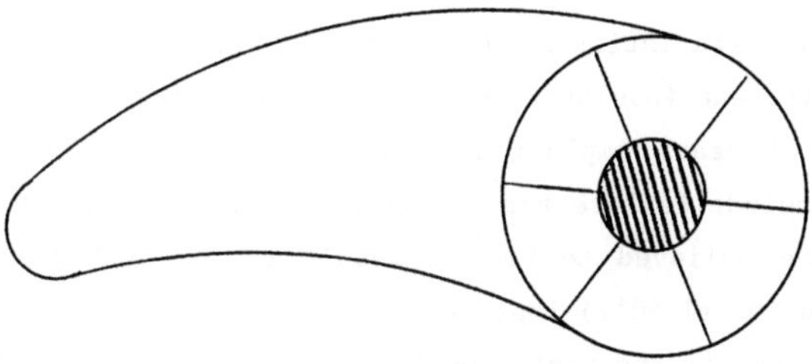

Figure 2: Cross-section of DBS wire showing 6 MP35N elements separated by silver, with a silver core and outer coating.

In late 1982, a degradation process was discovered occurring within polyurethane insulated leads. This was identified as an autoxidation mechanism, catalyzed by corrosion products of the conductor alloys. Initially, it was thought that the catalyst was a silver corrosion product.[55] Now, however, it is known that the catalyst is a cobalt species from the MP35N and that the corrosion of the MP35N was greatly accelerated by the presence of the silver.[72] The MIO mechanism is relatively well understood today, thus can be controlled.[72,73] Elimination of DBS in the second generation lead greatly reduced the incidence of MIO failure, but did not eliminate it. MP35N and Elgiloy still can corrode, albeit at much slower rates, and are equally succeptable to causing MIO.[72] The mechanical properties of MP35N and Elgiloy are still required for pacing leads, however. Thus, until a better conductor material is found, the application of a barrier coating to prevent corrosion so far appears to be highly effective in third and subsequent generation devices.

The Actuarial Performance of One Manufacturer's Polyether Polyurethane Insulated Cardiac Pacing Leads:

Only one device manufacturer (Medtronic, Inc.) reports the actuarial survival statistics of their post-market device surveillance.[74] Thus it is only possible to report objectively on that one company's experience. However, it cannot be implied that only this company's devices have failed (only that the other manufacturers do not report their device failures). Polyether polyurethane failure mechanisms affected the performance a few of this manufacturer's early pacemaker lead models significantly, caused a very low incidence of failure in some other models and had no significant effect in most others. It is unknown at this time what the failure statistics are for the rest of the industry.

The known objective statistics show that the actuarial survival of the first generation polyurethane insulated, bipolar, ventricular lead (Model 6972) was about 78% after 5 years implant and about 68% after 10.[74] While these statistics can include failure for a variety of reasons, almost all failures of this model are believed to be the result of ESC and MIO. The second generation device (Model 4012) implemented improved manufacturing techniques, some design enhancements and replacement of the DBS wire with MP35N. This has resulted in 95.8% actuarial survival in 5 years with trends projecting ≥90% survival in 10 years. However, most of the 4.2% failures over the 5 year period have been shown to be the result of conductor fracture due to another previously unknown mechanism called "subclavian crush" (caused by clamping between 2 bones, the first rib and the clavicle).[30] If only insulation failures are counted, the actuarial survival is 98.6% in 5 years, about 0.7% due to MIO and 0.7% due to ESC. The third generation device (Model 4004) has had 100% survival over a 2 year period. This is the result of conductor redesign to enhance the crush mechanism, education of implanters about the crush mechanism, further improvements in manufacturing processes and barrier coated MP35N wire. Based on rigorous accelerated tests that have been validated by real time studies and several years of canine testing, the performance of this device is projected to be much superior to the second generation device. A fourth generation lead (Model 4024) has just been approveg by the FDA. This lead model has even more improvements and is expected to have still better performance.

OTHER POLYETHER POLYURETHANES AND DEVICES

Polyurethane diaphragms have been tested extensively in artificial hearts.[75,76] One such system consists of two polyurethane blood pumps separated by 3 membranes into blood and energy transmission chambers[77]. The longest survival time in animals is over 200 days. Complications and problems are said to be due mostly to the material. Small lesions, cracks, and slight flaws or biodegradation are the initiation sites for early thrombosis and later calcification, which appears to by typical of polyurethanes used in this application.

The hydrolytic stability of microporous Mitrathane polyether polyurethane urea was evaluated at pH 7 & 9, at 37, 60 and 80° C for periods of up to 968 days[78]. The samples held at 80° C showed the greatest mechanical strength loss, which was attributed to hydrolysis of the urethane groups. The authors concluded, however that the gross in vivo surface degradation observed after 6 months could not have been caused by hydrolysis because the polymer was found to be (hydrolytically) stable at 37° C for at least 11 months.

The mineralization of polymers including polyether polyurethanes in many devices, especially heart valves and blood pump diaphragms is well known.[79] None-the-less, the development of polyurethane valves continues.[80,81] While the mechanism of mineralization is not yet completely understood, it is believed to be associated with the adsorption of low density lippoproteins, surface charge, the patient's age, flexation of the device's surfaces and cellular debris.[82,83,84,85]

CONCLUSIONS

There is no such thing as an implantable material per se. There are implantable devices made from materials which may or not be biostable in any given individual. What is truly important is the biostability of the device, which is a function of design-manufacturing process-material-tissue interactions. The degradation mechanisms that can affect an implanted polymer, include (but are not limited to) autoxidation, hydrolysis, mechanical mechanisms and mineralization. The catalysts and other causative factors that initiate and accelerate these mechanisms are extremely complex as well as numerous. There is no in vitro simulation for the hostile mammalian environment. Biostability, therefore, cannot be determined in vitro and reliance on in vitro testing should be viewed with skepticism. The qualification of a new material for use in an implanted device or a new use for an otherwise accepted material must include testing of the actual human use device, made by final manufacturing processes, for a long enough time (within reason) to assess its long term performance in a manner reasonably close to the final use before the device is marketed. Accelerated in vitro or in vivo tests may be acceptable to allow human clinical trials to begin for devices (such as pacemaker leads) that are generally not life sustaining, only if those tests have been validated by real time testing.

No matter how careful or thorough a manufacturer is, it is not possible to assure device reliability or even to know what the actual performance is over time without postmarket surveillance. Some failure mechanisms may not be discovered in animal tests, no matter how long they run. One example is crushing of pacemaker leads between the first rib & clavicle in humans, which cannot be predicted on the basis of the accepted canine model since canines do not have clavicles. It is certainly unrealistic to conduct animal tests for very long periods (for example to discover a failure mechanism that takes 10 years to occur). Thus, certain failure mechanisms may be species specific or take so long to manifest themselves that they escape detection in reasonably long term animal studies and even human clinical trials. No matter how biostable a device-material is, sooner or later, conditions may be right for device failure.

REFERENCES

1. Okita Y, Miki S, Kusuhara K, et al (1989) J Vasc Surg 10(5):554-556
2. Godard F, King M, Guidoin R, et al (1981) J Mal Vasc 6(3):167-171
3. Berger K and Sauvage LR (1981) Ann Surg 193(4):477-491
4. Zaikof GE (1985) JMS-Rev, Macromol Chem Phys C25(4): 551-597
5. Kopecek J and Ulbrich K (1983) Prog Polym Sci 9: 1-58
6. McClellan KA and Billson FA (1991) Ophthalmic Surg 22(2):74-77
7. Hayasaka S, Ishiguro SI, Shiono T, et al (1982) Am J Ophthalmol 93(1):111-117
8. Hudson JA and Crugnola A 1987) J Biomater Appl 1(4):487-501
9. Teoh SH, Martin RL, Lim SC, et al (1990) ASAIO Trans 36(3):417-421
10. Vondracek P and Dolzel B (1984) Biomaterials 5:209-214
11. The Pacemaker Telectronics Bulletin for the Cardiac Pacemaker Clinic. Telectronics, USA, January 1972, p7
12. Dow Corning Clean Grade Elastomers. Bulletin 51-209A, May 1979
13. In The International Plastics Selector, Cordura, LaJolla, p277, 1977
14. Dolezel B, Adamirova L, Naprstek Z, et al (1989) Biomaterials 10(2):96-100
15. Dolezel B, Naprstek Z and Lysenkova A (1979) Proceedings of the Sixth World Symposium on Cardiac Pacing. In C. Meere (ed) Montreal, PACESYMP, sect 36-5
16. Openheimer BS (1955) Cancer Res 15: 333

17. Herzenberg JE, Harrelson JM, Campbell DC et al (1988) Clin Orthop 228:88-93

18. Davidson JA, Gir S and Paul JP (1988) J Biomed Mater Res 22(3 Suppl):281-309

19. Remagen W and Morscher E (1984) Arch Orthop Trauma Surg 103(3):145-51

20. Eyerer P and Ke YC (1984) J Biomed Mater Res 18(9):1137-1151

21. Kurt M, Eyerer P, Ascherel R, et al (1988) J Biomater Appl 3(1):33-51

22. Wasserbauer R, Beranova M, Vancurova D et al (1990) Biomaterials 11(1):36-40

23. Beranova M, Wasserbauer R, Vancurova D, et al (1990) Biomaterials 11(7):521-524

24. Jongebloe WL, Figueras MJ, Humalda D, et al (1986) Doc Ophthalmol 61(3-4):303-312

25. Apple DJ, Craythorn JM, Olson RJ, et al (1984) Ophthamology 91(4):403-419

26. Mowbray SL, Chang SH and Dasella JF (1983) J Am Intraccul Implant Soc 9(2):143-147

27. Brase DW and Millender LH (1986) J Hand Surg 11(2):175-183

28. Boretos JW (1973) <u>In</u> Concise Guide to Biomedical Polymers, Charles C. Thomas, Publisher, Springfield, IL p11-12

29. Hartman LC, Bessette RE, Baier RE, et al (1988) J Biomed Mater Res 22(6):475-484

30. Bognolo DA (1985) In Barold SS (ed) Modern Cardiac Pacing, Futura Publishing Co, Mount Kisco, pp199-229

31. Blais P (1990) J Appl Biomater 1:197

32. Bloch B and Hastings G (1972) In Plastics Materials in Surgery, Thomas CC (Publisher) pp 97-98

33. Lelah MD and Cooper SL (1986) In Polyurethanes in Medicine, CRC Press, Boca Raton

34. Batich C, Williams J and King R (1989) J Biomed Mater Res 23(A3 Suppl):311-319

35. Szycher M and Siciciliano A (1991) J Biomater Appl 5:323-336

36. Weiss R (1991) Science 252:1060

37. Coury AJ, Stokes, KB, Cahalan PT, et al (1987) Life Support Syst 5(1):25-39

38. Stokes KB and Stephenson NL (1982) In Barold SS and Mujica J (eds) The Third Decade of Cardiac Pacing. Futura Publishing Co, Mount Kisco, pp365-416

39. Stokes K, Cobian K and Lathrop T (1979) In Meere C (ed) Proceedings of the VI World Symposium on Cardiac Pacing PACESYMP, Montreal, sect 28-2.

40. Stephenson NL (1980) Medtronic News X(2):10

41. Stephenson NL (1980) Medtronic News X(2):10

42. Stokes K and Cobian K (1982) Biomaterials 3: 225

43. Boretos JW, Pierce WS, Baier RE, et al (1975) J Biomed Mater Res 9:327.

44. Boretos JW (1972) J Biomed Mater Res 6:437

45. Silastic Medical Grade Elastomers Q-4635, Q7-4650, Q74665. Dow Corning, 1982

46. Ulrich H and Bonk HW (1982) Proceedings of the SPI 27th Annual Technical/Marketing Conferene p143

47. Stokes K (1982) Stimucoeur Medical 10:205

48. Byrd CL, McArthur, W, Stokes, K, et al (1983) PACE 6:686

49. Stokes K (1985) In Barold SS (ed) Modern Cardiac Pacing. Futura Publishing Co, Mount Kisco, pp173-198

50. Beyersdorf F and Kruezer J (1985) In Aubert D and Ector H (eds) Pacemaker Leads Elsevier, Amsterdam, 287-290

51. Stokes KB, Frazer WA and Carter EA (1984) Proceedings ANTEC 84 84:1073

52. Chawala AS, Blais P, Hinberg I and Johnson D (1988) Biomater Artif Cells Artif Organs 16(4):785-800

53. Szycher M (1988) J Biomater Appl 3(2):297-402

54. Stokes K, Urbanski P and Cobian K (1987) In Planck H, Syre I, Dauner M, et al (eds) Polyurethanes in Biomedical Enginering II. Elsevier, Amsterdam, pp109-127

55. Coury A, Cahalan P, Schultz E, et al (1984) Transactions Second World Congress on Biomaterials p252

56. Coury A, Cahalan P, Slaikeu P, et al (1985) Transactions 11th Annual Meeting of the Society for Biomaterials p134

57. Brauhmstedt-Sutherland K (July, 1991) Masters Thesis, Dept of Eng, U Minn

58. Stokes K (1988) J Biomater Appl 3:228-258

59. Zhao Q, Agger MP, Fitzpatrick M, et al (1990) J Biomed Mater Res 24:621-637

60. Zhao Q, Topham N, Anderson JM, et al (1991) J Biomed Mater Res 25:177-183

61. Sibille Y and Reynolds HV (1990) Am Rev Respir Dis 141:471-501

62. Anderson, JM, personal communication

63. Ratner BD, Gladhil KW and Horbett TA (1988) J Biomed Mater Res 22:509-527

64. Marchant RE, Anderson JM, Castillo E, et al (1986) J Biomed Mater Res 20:153-168

65. Smith R, Williams DF and Oliver C (1987) J Biomed Mater Res 21:1149-1166

66. Thoma RJ and Phillips RE (1987) J Biomed Mater Res 21(4):525-530

67. Thoms RJ (1987) J Biomater Appl 1(4):449-446

68. Klebanoff SJ (1968) J Bacteriol 95:2131-2138

69. Reynolds HY (1985) Dis Mon 31:1-98

70. Roos D and Balm AJM (1980) In Sbarra AJ and Strauss RR (eds) The reticuloendothelial system. A Comprehensible treatise: biochemistry and metabolism. Plenum, New York pp189-299

71. Stokes KB and Anderson JM (1991) PACE 14(4):731

72. Stokes, K, Urbanski, P and Upton J (1990) J Biomater Sci Polymer Edn 1(3):207-230

73. Stokes KB, Coury A and Urbanski P (1987) J biomaterials Applications 1:411-448

74. Medtronic Product Performance Report, Medtronic Inc, 7000 Central Ave N.E., MS T152, Minneapolis, MN 55432, USA

75. Phillips W, Pierce W, Rosenberg G, et al (1980) In Szycher M and Robinson E (eds) Synthetic Biomedical Polymers Concepts and Applications. Technomic, Westport, pp39-57

76. Olsen D (1987) Transactions of the 13th Annual Meeting of the Society for Biomaterials p168

77. Bucherl ES (1985) Z Kardiol 74 (Suppl 6):65-71

78. Paynter RW, Askill IN, Glick SH, et al (1988) J Biomed Mater Res 22(8):687-698

79. Hilbert SL, Ferrans VJ, Tomita Y, et al (1987) J Thorac Cardiovasc Surg 94(3):419-429

80. Yu LS, Yuan B, Bishop, D, et al (1989) ASAIO Trans 35(3):301-304

81. Lo HB, Herold M, Reul H (1988) ASAIO Trans 34(3):839-844

82. Dong DE, Andrade JD & Coleman DL (1987) J Biomed Mater Res 21(6):683-700

83. Stupp SI & Kusleika R (1985) J Biomed Mater Res 19(3):321-324

84. Imai Y (1985) Artif Organs 9(3):255-258

85. Schoen FJ, Harasaki H, Kim KM, et al (1988) J Biomed Mater Res 22(A1):11-36

A Comparison of *In Vivo* Degradation of Novel Polyurethanes with Performance in *In Vitro* Accelerated Tests

[1]Brandwood A., [1]Noble K.R., [1]Schindhelm K., [2]Meijs G.F., [2]Gunatillake P.A., [2]Chatelier R.C., [2]McCarthy S.J. and [2]Rizzardo E.

[1]Centre for Biomedical Engineering, University of New South Wales
Kensington, NSW, Australia 2033

[2]CSIRO Division of Chemicals and Polymers
Private Bag 10, Clayton, Vic, Australia 3168

ABSTRACT

In *in vitro* accelerated tests, both novel and commercially produced polyurethanes were subjected to severe hydrolysing and oxidising conditions for 24 hours. Degradation in these tests was compared to degradation of the same materials in an ovine *in vivo* accelerated test in which polyurethane dumbbells were stretched over poly(methyl methacrylate) holders to a constant strain and implanted for 90 days subcutaneously. Further samples of the same materials were implanted unstressed for 6 months. Degradation in both *in vitro* and *in vivo* accelerated tests was correlated with degradation observed in the 6 month *in vivo* study. The *in vitro* test is suitable for initial screening of novel polyurethanes. The accelerated *in vivo* model generated environmental stress cracking in 90 days and was able to discriminate between materials of differing biostability. Three of the novel materials studied were more biostable than the commercial materials. The accelerated *in vivo* model provides a technique to identify materials which are more biostable than existing commercial polyurethanes.

INTRODUCTION

Polyurethanes are used in a wide range of medical devices in applications as diverse as wound dressings, sutures and ventricular assists. These materials combine excellent mechanical properties (high strength, toughness and extensibility) with low thrombogenicity. This combination of properties makes them particularly suitable for chronic blood contacting applications. Unfortunately, polyurethanes when used in chronically implantable devices tend to undergo degradation resulting in surface cracking[1]. The precise causes of the cracking are not fully understood but a range of mechanisms has been proposed including non specific and enzymatic oxidative and hydrolytic processes, metal ion induced oxidation and the action of free radicals. A common factor is that the observed surface degradation is promoted by mechanical stress. The phenomenon is often referred to as *stress cracking*. The occurrence of stress cracking has restricted the use of biomedical polyurethanes in chronic applications.

Stress cracking *in vivo* develops slowly, typically over periods of many months. This presents a problem in the development of novel, more biostable materials in that *in vivo* experimental evaluations are necessarily lengthy. There is a requirement for experimental techniques which can give more rapid indications of biostability - in particular resistance to stress cracking.

H. Planck M. Dauner M. Renardy (Eds.)
Degradation Phenomena on Polymeric Biomaterials

One approach is to expose novel materials *in vitro* to severe hydrolytic or oxidative conditions, in order to obtain accelerated degradation. It has been pointed out[2] that such techniques may induce degradation by mechanisms (for example thermal degradation) that do not occur *in vivo*. This may limit the use of chemical tests as predictors of *in vivo* performance.

An alternative approach is to apply external stress to materials while they are implanted in tissue. This application of stress would be expected to accelerate the formation of stress cracking.

We have carried out a series of *in vitro* accelerated tests in which novel polyurethanes, along with commercially produced materials, have been subjected to severe hydrolysing and oxidising conditions for periods of 24 hours. The degradation in these tests was compared to the observed degradation of the same materials in an ovine *in vivo* accelerated test model in which dumbbell specimens, prepared from flat sheets of polymer, are stretched over poly(methyl methacrylate) (PMMA) holders to a constant strain and implanted for 90 days subcutaneously in sheep. The model allows comparison of biodegradation of stressed materials with implanted unstressed controls. This model has the advantages that it requires only minimal quantities of material and that specimen preparation is simple and straightforward. Also, since test pieces are prepared from sheet material, then either moulded or solvent cast materials may be evaluated. In a third experiment, further samples of the same materials were implanted unstressed for 6 months. In this way we were able to compare the results of *in vitro* and *in vivo* accelerated tests with longer term *in vivo* degradation and thus assess the validity of the accelerated tests as predictors of *in vivo* degradation.

MATERIALS AND METHODS

Materials Preparation

Eight novel polyetherurethanes A to H were formed into flat sheets of nominal dimensions 60 mm x 100 mm x 0.5 mm, by either compression moulding or solvent casting. Compression moulding was carried out at nominal pressures of 8-12 tons and temperatures of 100°-200^oC. Solvent cast sheets were dried for 7 days under a flow of dry nitrogen at 40^oC. Test specimens were cut from the flat sheets using specially manufactured dies.

In Vitro *Degradation.*

Dumbbell specimens of polyetherurethane were enclosed in a PTFE sheet that was bent once and tied loosely with PTFE tape. The PTFE sheet had holes punched in it to allow circulation of test solution. This arrangement weighted the sample so that it did not float and prevented test pieces from adhering to each other. The enveloping was sufficiently loose that exposure of the polyurethane to the reagent solution was unhindered.

Replicate specimens (n=5) were exposed for 24 hours to one of the following treatments: reflux in 2M HCl, 5M NaOH, 25% hydrogen peroxide or sodium hypochlorite (4% available chlorine); water at 120°C in a glass lined pressure vessel. As a control, specimens were also refluxed in de-ionised water without any of the oxidising or hydrolysing reagents present.

After 24 hours, the specimens were retrieved and tensile properties were determined, following the procedures described in ASTM-D412-87, using an Instron tensometer. Tensile properties of control, untreated specimens were also determined.

In Vivo *Accelerated Test*

Dumbbell specimens (n=6) (Figure 1a) were cut from the sheet materials using a specially manufactured die. They were stretched over poly(methyl methacrylate) holders (Figure 1b,c). In this configuration the central portion was extended to 150% strain. A 3-0 polypropylene suture ligature was tied firmly around the centre of the dumbbell (Figure 1c) to act as a stress raiser. Control, unstressed specimens were loosely attached to PMMA holders by two polypropylene sutures (Figure 1d). The different types of material were identified by a binary code marked on the PMMA holders.

All specimens were sterilised by ethylene oxide followed by degassing for 7 days. Specimens were then implanted sub-cutaneously in the dorsal thoraco-lumbar region of 1-2 year old crossbred wether sheep.

Two further sets of stressed and unstressed specimens were prepared and sterilised as above. One set was stored in darkness in air at room temperature while the other set was stored in sterile phosphate buffered saline (PBS) *p*H 7.2 at 37°C in darkness, for 90 days.

All specimens were retrieved after 90 days. Surrounding fibrous encapsulation was dissected away and the specimens were cleaned by rinsing in deionised water followed by soaking in 0.1M NaOH for 4 days then rinsing again in deionised water and air drying. Specimens stored in air or PBS were similarly treated.

All specimens including non-implanted controls were then prepared for examination by scanning electron microscopy (SEM).

Specimens were examined at magnifications up to 2000. The extent of surface degradation of each stressed, explanted dumbbell was ranked on a scale of 1 (no evidence of surface degradation, surface smooth) to 6 (extensive, deep stress cracking with significant loss of material from the surface). Any surface features also observed on the non-implanted, dry stored controls were ignored when determining ranking. Any surface degradation on unstressed, implanted controls was also recorded.

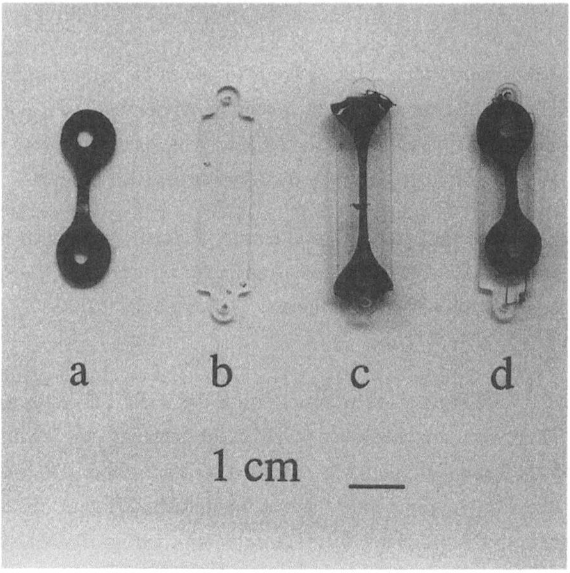

Figure 1. Test pieces: a, polyurethane dumbbell; b, PMMA holder; c, stressed implant; d, unstressed implant.

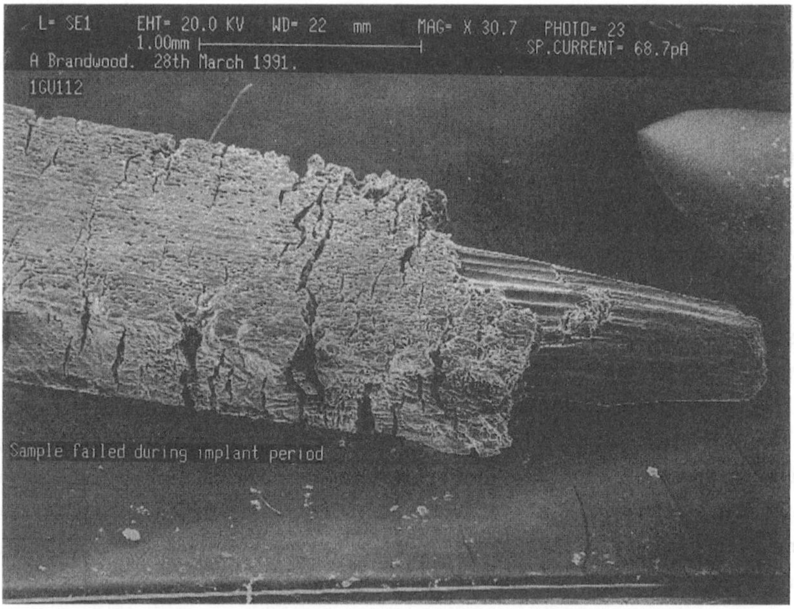

Figure 2. Micrographs showing different levels of surface degradation. Polymer H.

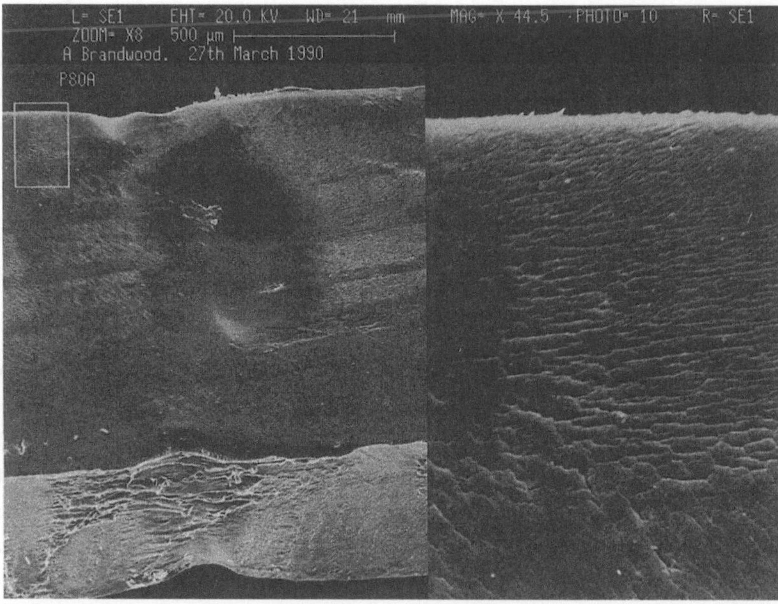

Figure 2 (continued) Pellethane™ 2363-80A.

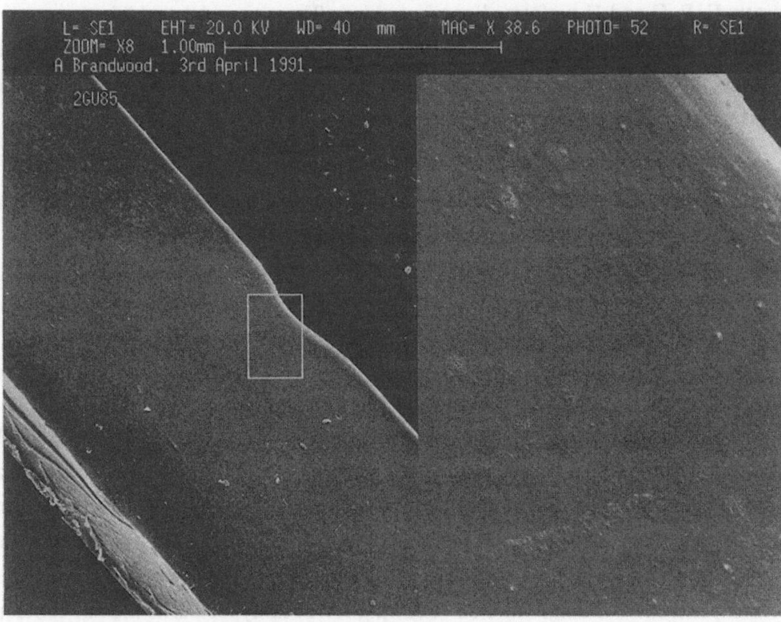

Figure 2 (continued) Polymer A.

In vivo *unstressed implants*

Rectangular specimens 10 x 35mm were cut from polyurethane sheets (these sheets were prepared with a nominal thickness of 1 mm rather than 0.5 mm). The specimens were identified by a binary code punched into the specimen using a modified 19 gauge needle. Specimens were sterilised by ethylene oxide and implanted sub-cutaneously as described above. After a period of 180 days, all specimens were explanted and cleaned as above. The specimens were then examined by SEM and the surface degradation again ranked on a scale 1 (no degradation) to 6 (extensive stress cracking), as described above.

RESULTS AND DISCUSSION

The *in vivo* experiments showed a wide range of surface degradation (Figure 2) with the commercial materials falling in the middle of the range. Note that Polymer D was an equivalent formulation to Pellethane 2363-80A which was prepared using the same laboratory scale techniques as the novel polymers. This material was prepared as a control to confirm that laboratory scale synthesis did not have a deleterious effect on properties when compared to similar formulations produced on a commercial scale. Polymer D performed very similarly to Pellethane 2363-80A in the *in vivo* experiments: both materials showed stress cracking concentrated around the suture ligature stress raiser in the stressed dumbbells, with patches of stress cracking elsewhere on the sample.

Polymers E, F, G and H performed poorly *in vivo*. These materials exhibited extensive stress cracking in both *in vivo* studies. In the 90 day, stressed study, several samples failed during implant, at the site of the suture ligature. Stress cracking was also observed after 90 days on implanted unstressed samples of these materials.

In contrast, Polymers A and B, (which are of different chemical composition to polymers E to H) showed no evidence of surface degradation in either *in vivo* experiment.

For all materials, the stressed and unstressed control specimens which were stored in air and in PBS at 37°C for 90 days showed no evidence of surface degradation when examined by SEM.

The results of *in vitro* accelerated tests are shown in Table 1. Figures for ultimate tensile stress (UTS) are given. Other tensile parameters showed similar trends. Table 2 compares the ranking obtained from the *in vivo* experiments with the UTS results for the degraded materials expressed as a percentage of the UTS of the same material refluxed in de-ionised water alone. Note that the polymers in Tables 1 and 2 are ranked in order to their biostability as seen in the *in vivo* experiments. Changes in UTS following reflux in hypochlorite are not correlated with *in vivo* biostability. However, the other accelerated tests - both the hydrolytic tests and reflux in peroxide - show a correlation with *in vivo* results. It is surprising that hydrolytic degradation

Table 1. Ultimate tensile stress following accelerated *in vitro* degradation.

Polymer	UTS (MPa) after treatment in:						
	Unt.	*Water*	*120C*	*HCl*	*NaOH*	H_2O_2	*NaOCl*
A	20.0	19.0	19.3	20.5	17.1	18.6	0.0
B	13.0	13.5	10.8	11.0	13.0	12.0	10.5
C	19.0	20.0	15.7	17.0	18.0	16.5	12.0
D	10.5	9.5	7.4	8.3	8.0	7.7	0.0
P80A	29.7	29.2	5.5	13.6	18.7	15.7	19.9
Biomer	30.8	29.8	17.3	4.6	13.0		20.1
E	10.6	0.0	0.0	0.0	0.0	0.0	0.0
F	23.2	24.0	9.8	0.0	21.0	5.4	14.0
G	17.0	14.3	4.1	0.0	14.6	0.0	4.2
H	26.6	3.4	0.0	0.0	0.0	0.0	0.0

P80A, Pellethane™ 2363-80A; *UTS,* ultimate tensile stress; S_{90}, 90 day, stressed implant; U_{180}, 180 day, unstressed implant; *Unt.*, untreated; *Water,* de-ionised water; *120C,* water at 120°C in a pressure vessel.

Table 2. Results of *in vivo experiments* and results of accelerated tests expressed as a percentage of the mean UTS of the water refluxed control material.

Polymer	*In Vivo* Ranking		*UTS* as % of Water Refluxed Control				
	S_{90}	U_{180}	*120C*	*HCl*	*NaOH*	H_2O_2	*NaOCl*
A	1	1	102	108	90	98	0
B	1	1	80	81	96	89	78
C	1	2	79	85	90	83	60
D	2.5	4	78	87	84	81	0
P80A	2.5	3	19	47	64	54	68
Biomer	4.5	4	58	15	44	0	67
E	6	4	0	0	0	0	0
F	5	6	41	0	88	23	58
G	6	5.5	29	0	102	0	29
H	6	5	0	0	0	0	0

shows a correlation with *in vivo* stability. The polymers studied were all polyetherurethanes, a class of polyurethanes which are believed to be relatively hydrolytically stable although prone to oxidation[1].

In conclusion, notwithstanding concerns about degradation mechanisms[2], the *in vitro* accelerated test model does have some value as an initial screen of novel biomedical polyetherurethanes. The test is rapid and simple. It appears to be conservative - all of the materials which showed high biostability *in vivo* also showed high stability in the *in vitro* test. Nonetheless, such tests are necessarily crude and their use can only be restricted to an initial screening process.

In contrast the accelerated ovine *in vivo* model will generate environmental stress cracking in 90 days and is able to discriminate between materials of differing biostability. In particular, the model is able to identify materials which have higher biostability than existing commercial materials.

ACKNOWLEDGMENTS

Support was provided under the Generic Technology component of the Industry Research and Development Act in conjunction with Cyanamid Pty Ltd, Terumo Pty Ltd and Telectronics Pty Ltd.

The technical assistance of Donna McIntosh, Edith Cheong and Veronika Tatarinoff is gratefully acknowledged.

REFERENCES

1. Szycher M (1991) In: Sharma CP, Szycher M (eds) *Blood Compatible Materials and Devices*. Technomic: Lancaster, Penn. p33-86

2. Hum LL, MacGregor DC, Mayer DW, Pinchuk L (1990) *J Appl Biomat* **1**:193

SELECTED DISCUSSION REMARKS:

Stokes: You may have polymers, that will survive your test. In our tests we had polymers, that were not stress-cracked, but they still can go down because of metal-catalyzed oxidative mechanisms.

Brandwood: Indeed. We haven't looked on metal ions with this particular technique.

ACCELERATED DEGRADATION TEST ON RESORBABLE POLYMERS

Berthold Buchholz

Boehringer Ingelheim KG, A Entwicklung Chemikalien
D-6507 Ingelheim, Germany

INTRODUCTION

Polyesters derived from aliphatic hydroxy carbonic acids are a well known class of synthetic bioresorbable polymers. The biodegradation normally proceeds by random bulk hydrolysis of ester bonds without a significant contribution of enzymatic attack. Enzymatic surface erosion is restricted to rubbery polylactones with low elastic moduli [1], with only one example reported in the literature [2]. As rubbery polylactones, rubbery poly(ethylene carbonate) [3] and poly(trimethylene carbonate) [4] are reported to degrade in vivo by enzymatic action.

The random bulk hydrolysis of polylactones results in an immediate decrease of molecular weight after contact with water and, after an extended period of time, in weight loss by diffusion of oligomers, generated by hydrolysis [5]. The degradation rate depends on the chemical structure, but also on experimental conditions, like sample preparation, pH, temperature, etc. Under physiological conditions, complete degradation can last up to several years.

This paper describes initial results of accelerated hydrolysis tests at elevated temperature (80 °C). Although accelerated hydrolysis does not allow a precise calculation of the hydrolysis rate at physiological temperature, this procedure is to enable a comparison of different polymers, in particular to evaluate the degradability of novel materials. The polymers under investigation are poly(L-lactide) (P-LLA), poly(DL-lactide) (P-DLLA), and a series of rubbery copolymers comprising DL-lactide and trimethylene carbonate units (P-DLLA-TMC).

H. Planck M. Dauner M. Renardy (Eds.)
Degradation Phenomena on Polymeric Biomaterials

EXPERIMENTAL

Materials

P-LLA and P-DLLA are Resomer[R] from Boehringer Ingelheim. Synthesis and properties of P-DLLA-TMC are described elsewhere [6]. The composition of the copolymers is given in percentage by weight.

Sample Preparation

P-LLA and P-DLLA were compressed to tablets of 200 mg weight and 13 mm diameter. To obtain a reproducible crystallinity, P-LLA was reprecipitated from chloroform / petroleum ether and dried at 80 °C prior to compression. For the rubbery P-DLLA-TMC, samples of 200 mg weight and about the same shape as the tablets were cut from the bulk.

Hydrolysis Tests

Polymer samples were immersed in 150 ml phosphate buffer of pH 6.1 at 80 °C under magnetic stirring. The specimens were periodically removed, vacuum dried at 40 °C, and subjected to viscosity measurement. The inherent viscosity (IV) of all polymer samples was measured in chloroform solution (0.1 %, 25 °C) using an automated Ubbelohde viscometer. Repeated hydrolysis tests showed a good reproducibility with a standard deviation of below 0.1 dl/g (according to a variation coefficient of below 7 %). Weight loss was measured only with the crystalline P-LLA. The test conditions do not permit determination of weight loss of the amorphous polymers, since form stability at 80 °C is low, and a total recovery of degraded samples from the buffer solution is not possible. A comparative test series at 37 °C was performed with P-DLLA.

RESULTS AND DISCUSSION

Hydrolysis Kinetics

As an indication for changes in molecular weight during hydrolysis, the inherent viscosity (IV) of degraded samples is measured as a function of time. Initial IV of the different polymers and changes in IV are reported in table 1.

Table 1
Inherent Viscosity vs. Degradation Time

Time [h]	Inherent Viscosity [dl/g] P-LLA		P-DLLA	P-DLLA-TMC 90/10	70/30	50/50	30/70	10/90	P-DLLA 37 °C	Time [h]
0	4.14	7.78	1.07	1.50	1.11	1.06	1.05	1.24	1.07	0
1			0.96	1.48	1.09					1
2	4.00	7.59	0.86	1.43	1.08					2
4	3.91	7.39	0.69	1.28	1.02					4
6	3.69	7.23	0.52							6
8	3.67	7.02	0.44	1.02	0.92					8
10			0.36	0.96	0.80	0.88				10
14			0.24	0.73	0.71					14
16	3.17	6.56	0.23	0.59	0.67	0.81	0.95			16
20			0.16	0.56	0.59					20
24	2.75	5.99	0.13	0.32	0.51	0.75	0.92	1.19		24
26			0.10							26
32				0.26	0.34					32
40			0.09	0.16	0.25	0.50	0.85			40
48	1.79			0.10	0.19	0.43	0.78	1.18	1.06	48
49		4.55								49
62						0.28	0.65			62
72					0.10	0.26	0.62	1.16	1.05	72
86							0.45			86
96	0.65	2.38				0.14	0.46	1.08		96
120							0.30	1.08		120
144	0.25	1.02					0.21	1.00	1.04	144
168									1.01	168
192	0.19	0.45						0.89	1.00	192
240								0.74		240
288								0.67	0.98	288
336								0.50	0.95	336
408									0.91	408
432								0.36		432
504								0.23	0.88	504
552								0.21		552
673									0.83	673
840									0.76	840
1008									0.73	1008
1344									0.63	1344
1680									0.57	1680
2352									0.43	2352
2520									0.42	2520
3096									0.35	3096

The degradation profile of all polymers under investigation is qualitatively similar, with a rapid decrease of IV in the initial stage of hydrolysis and smaller changes at later periods.

For the kinetics of polyester degradation in the solid state two equations were derived, one for an autocatalytic reaction and one for uncatalyzed hydrolysis [7]. If the hydrolysis is autocatalyzed by generated carboxylic end groups, the following rate law applies:

$$d\,[COOH] / d\,t = k_1 \cdot [COOH] \cdot [ester] \cdot [H_2O] \tag{1}$$

Assuming [ester] and [water] = constant and [COOH] ~ $1/M_N$ equation 1 simplifies to:

$$d\,(1/M_N) / d\,t = k_2 \cdot 1/M_N \tag{2}$$

Integration gives:

$$\ln M_N = \ln M_N{}^{\circ} - k_2 \cdot t \tag{3}$$

If changes in polydispersity are neglected, the decrease of intrinsic viscosity [eta] during polymer degradation can be described by substitution of M_N (number average molecular weight) for M_{vis} (viscosity average molecular weight) of the Mark-Houwink equation (4):

$$[\eta] = k \cdot (M_{vis})^{\alpha} \tag{4}$$

$$\ln [\eta] = \ln [\eta]^{\circ} - k_3 \cdot t \quad \text{with } k_3 = k_2 \cdot \alpha \tag{5}$$

For uncatalyzed hydrolysis equations 6 and 7 apply:

$$d\,[COOH] / d\,t = k_4 \cdot [ester] \cdot [H_2O] \tag{6}$$

$$(\,[\eta]^{\circ} / [\eta]\,)^{1/\alpha} = 1 + k_5 \cdot t \tag{7}$$

Disregarding the minor differences between intrinsic and inherent viscosity, the IV data from table 1 are analysed according to eq. 5 and 7. Mark-Houwink exponents are taken from literature: P-LLA: alpha = 0.82 [8]; P-DLLA: alpha = 0.64 [9]. The hydrolysis of P-DLLA-TMC is analysed according to eq. 5 only, as no Mark-Houwink parameters are available.

Table 2
Hydrolysis Rate at 80 °C

Polymer	HL [h] (eq 5)	k3 [1/h]	k2 [1/h]	R (eq 5)	R' (eq 7)
P-LLA IV 7.8 dl/g	50	0.014	0.017	0.997	0.893
P-LLA IV 4.1 dl/g	35	0.020	0.024	0.999	0.961
P-DLLA	8	0.089	0.139	0.994	0.929
P-DLLA-TMC 90/10	12	0.058		0.996	
P-DLLA-TMC 70/30	20	0.036		0.996	
P-DLLA-TMC 50/50	33	0.021		0.996	
P-DLLA-TMC 30/70	61	0.0113		0.981	
P-DLLA-TMC 10/90	209	0.0033		0.983	
P-DLLA 37 °C	1800	0.00039	0.0006	0.999	0.975

The poor correlation coefficients (R', table 2) and visual inspection of the corresponding plots show the failure of the uncatalyzed model. On the other hand, the correlation (R, table 2) for the autocatalyzed model shows good approximation to the experimental data. This result is obtained, although the first-order model does not consider changes in polydispersity and water uptake during degradation and although the small difference between intrinsic and inherent viscosity is disregarded. Selected semilog plots consistent with eq. 5 are displayed in figure 1 . From the linear regression of log (IV) vs. time the half life (HL) and rate constant (k_3) for the decrease of IV, as well as the rate constant (K_2) for the decrease of M_{vis} are calculated. The regressions generally correspond with the experimental data within an error of below 0.1 dl/g.

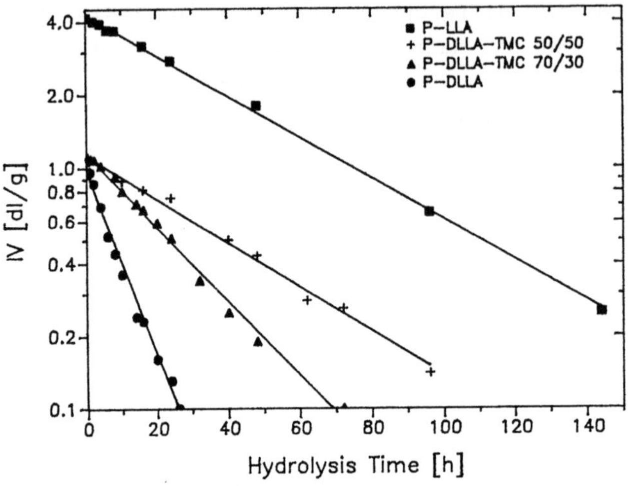

Figure 1
log IV vs. Hydrolysis Time

Some deviations from first-order kinetics are found, which reflect the limitations of the relatively simple model:

1. Comparing the rate constants of P-LLA with initial IV of 4.1 and 7.8 dl/g, a difference of 29 % is found. According to the model both polymers should be hydrolyzed with the same rate. This comparison shows that the model is not perfect and begins to fail when polymers with extremely different initial molecular weights are compared. This might reflect that the impact of changes in polydispersity and water uptake during degradation is less important when investigating polymers with comparably low molecular weights, but becomes more and more important as the initial molecular weight increases.

2. In the very low viscosity range of P-LLA and P-DLLA below ca. 0.1 - 0.2 dl/g experimental IV data are higher than calculated. This indicates that eq. 5 loses its relevance in this viscosity range. The equation will hold only until loss of mono- and oligomers reduces the concentration of catalytic active end groups [7]. In the experiments with P-LLA the corresponding hydrolysis time coincides with the onset of weight loss.

3. For P-DLLA-TMC with high TMC content the autocatalytic model does not seem to correlate with the experimental results as well as for the other polymers, since the semilog plots show some deviation from a straight line and correlation coefficients for these polymers are lower. Starting out from the results and discussions of Pitt and coworkers [4] who evaluated the hydrolysis of poly(trimethylene carbonate) at 37 °C it can be suggested that this deviation from first-order kinetics is attributed to low concentration of catalytic active end groups in these carbonate rich copolymers. Compared with carboxylic end groups, the end group generated by hydrolysis of carbonate bonds has no catalytic action, since the carbonic acid end group is a weaker acid and is unstable.

Temperature Dependency

By comparing the hydrolysis rates of P-DLLA at 80 °C and 37 °C, an apparent activation energy (E_{act}) of 27.5 Kcal/mole is calculated, assuming that temperature dependency follows the Arrhenius law (8).

$$\ln k = A + (E_{act} / R \cdot T) \qquad (8)$$

This value is in accordance with an activation energy reported in the literature: Mason et al. [10] calculated 29.6 Kcal/mole from hydrolysis experiments with P-DLLA at 25 °C and 37 °C. Makino et al. [11] obtained a different result: they reported an activation energy of 19.9 Kcal/mole at temperatures between 21 °C and 45 °C, however, this value is based on rate constants obtained from changes in <u>weight</u> average molecular weight from GPC measurements.

The good agreement between E_{act} obtained in this study and E_{act} reported at lower temperature [10] might suggest that the Arrhenius law is valid in the temperature range under investigation, although it covers the glass transition of P-DLLA (57 °C). In contrast, temperature coefficient of hydrolysis rate of poly(glycolide) is reported to show a discontinuity at the glass transition temperature [12]. Thus, further experiments are necessary to investigate whether the glass transition of P-DLLA affects the temperature coefficient of hydrolysis, since the present result is based on a double point measurement only. In addition, it has still to be investigated whether the temperature dependency of P-DLLA also applies to polymers with different glass transition and to semicrystalline polymers. This information is required for using results from accelerated tests for calculating hydrolysis rates at 37 °C.

CONCLUSIONS

Accelerated hydrolysis tests on resorbable polyesters at 80 °C are described. Changes of IV vs. time are in good agreement with a kinetic model, based on autocatalytic action of generated carboxylic end groups. Kinetics follow the first-order law over a wide viscosity range. Comparing the hydrolysis of P-LLA with considerably different initial IV, results in a divergence of about 30 %. Deviations from first order kinetics are found at the end period of degradation when weight loss of samples starts, and in carbonate rich copolymers. The hydrolysis tests at elevated temperature are useful for initial evaluations of novel materials and allow for a comparison of different polymers: in the series of P-DLLA-TMC the influence of copolymer composition on degradation rate is clearly demonstrated.

For calculations of hydrolysis rates at physiological temperature from data obtained at 80 °C, further investigations are required to study the potential impact of glass transition and crystallinity on temperature coefficient.

Since the test conditions do not allow for investigating enzymatic actions, the significance of test results for <u>bio</u>degradation is in general restricted to those polymers which are subject to random bulk hydrolysis in vivo. As discussed for poly(trimethylene carbonate) [4], the importance of enzymatic erosion might become a relevant factor in P-DLLA-TMC with high TMC content. As a result, the large differences found in the various copolymers might be to some extend expected to level under in vivo conditions by changing the degradation mechanism.

REFERENCES

1. A. Schindler, C. G. Pitt, Transact. Soc. Biomat. 8, 217, 1985
2. C. G. Pitt, R. W. Hendren, A. Schindler, J. Controlled Rel. 1, 3-14, 1984
3. T. Kawaguchi, M. Nakano, K. Juni, S. Inoue, Y. Yoshida, Chem. Pharm. Bull. 31, 1400-1403, 1983
4. K. J. Zhu, R. W. Hendren, K. Jensen, C. G. Pitt, Macromolecules, 24, 1736-1740, 1991
5. F. G. Hutchinson, B. J. A. Furr, J. Controlled Rel. 13, 279-294, 1990
6. B. Buchholz, Transact. Soc. Biomat. 14, 221, 1991
7. C. G. Pitt, Z. W. Gu, J. Controlled Rel. 4, 283-292, 1987
8. E. W. Fischer, H. J. Sterzel, G. Wegner, Kolloid-Z. u. Z. Polymere, 251, 980-990, 1973
9. J. Rak, J. L. Ford, C. Rostron, V. Walters, Pharm. Acta Helv. 60, 162-169, 1985
10. N. S. Mason, C. S. Miles, R. E. Sparks, Coat. Plast. Chem. 42, 436-440, 1980
11. K. Makino, M. Arakawa, T. Kondo, Chem. Pharm. Bull. 33, 1195-1201, 1984
12. A. M. Reed, D. K. Gilding, Polymer, 22, 494-498, 1981

SELECTED DISCUSSION REMARKS:

Van den Berg: Can you explain the autocatalytic mechanism in buffered solution?

Buchholz: The explanation for the autocatalytic mechanism in buffered solution is given in a paper of Pitt: It is not the action of carboxyl groups which are leached out by diffusion of oligomers or even monomers, because they are neutralized by the buffer. It is the catalytic action of carboxylic end groups which are still in the polymer bulk and are not leached out.

Kricheldorf: Why did you select a temperature of 80°C?

Buchholz: For good acceleration you need high temperature. You have to go over the glass transition point of poly-DL-lactide, which is around 55°C. The highest temperature, which is practical in laboratory work, is 100°C, but because of the loss of water this temperature is not the best. So we chose 80°C, it might also have worked with 90°C.

Nakamura: From my experience it is not useful to consider high temperature degradation for simulation of in vivo degradation, because the degradation behaviour is quite different below the glass transition temperature and above. But I agree with you, for the preliminary evaluation of the degradation of a polymer, it is very practical to save experimental time.

Buchholz: I completely agree with you.

KINETICS OF RESORPTION OF DIFFERENT SUTURE MATERIALS DEPENDING ON THE IMPLANTATION SITE AND THE SPECIES

M.A. Scherer, H.-J. Früh, R. Ascherl*, H. Mau, W. Siebels, G. Blümel

Institut für Experimentelle Chirurgie der TU-München
*Orthopädische Klinik und Poliklinik r.d.I. der TU-München
Ismaninger Str. 22, 8000 München 80, Germany

SUMMARY

In this experimental series in sheep and rats, a strict correlation between the resorption kinetics of suture material made of polyhydroxybutyrate (PHB), polyglactin 910 (PLA), polyglycol-trimethyl-carbon (PGA/C) and polydioxanone (PDS) and the site of implantation as well as the species was demonstrated. Maximum observation period was 6 to 12 weeks, a total of 361 samples underwent biomechanical testing.

Clinical consequences: The intraarticular environment leads to a faster rate of resorption, thus endangering the results of intraarticular ligament or meniscus surgery: The surgeon relies on the data given in the literature - and most of these are derived from experiments with subcutaneous implants.

Keywords: suture; resorption; degradation; experimental.

INTRODUCTION

A lot of work has been done with the scope to know determinant factors that affect the resorption and/or degradation of polymeric biomaterials. VERT (22) listed those that were held responsible for alterations of aliphatic polyesters (see Tab. 1).

Tab. 1: Factors that may affect resorption/ degradation of aliphatic polyesters

chemical structure
configurational structure
molecular weight distribution
presence of low molecular weight compounds
shape
sterilizing process
morphology
annealing
storage history
site of implantation
adsorbed and absorbed compounds in vivo
physico-chemical factors
hydrolysis
stresses

mod. acc. to Vert M and Guerin Ph, 1991 (22)

Experimental & Clinical Background

According to the experience of our group, resorbable drug carriers (collagen-antibiotic compounds)

H. Planck M. Dauner M. Renardy (Eds.)
Degradation Phenomena on Polymeric Biomaterials

show different release kinetics and bioavailability of the incorporated drug, when they are implanted into different sites of the body, along with varying states of inflammation and in several species. Fig. 1 gives the mean aminoglycoside serum levels after implantation of 10 mg of gentamicin-collagen/kg BW in the rat, the rabbit and as a therapeutic application in man. There is a striking difference concerning the serum levels and thus the elution rate of the drug from the carrier. Several causes may be taken into consideration, among these are the body temperature, basal metabolism, implantation volume per body area, state of inflammation and vascularity.

Fig. 1: Gentamicin serum levels after application of 10 mg/kg bw in rabbits, rats and men. Bioavailability of the drug is depending on the species.

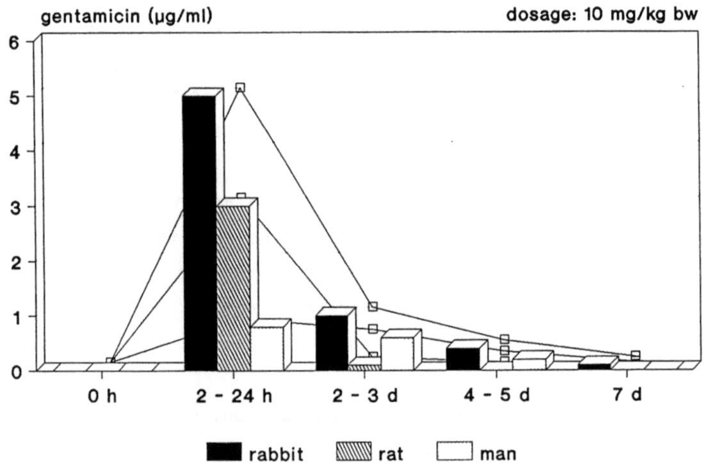

Surgery of the knee: There is a longstanding controversy in the clinical literature whether an isolated ACL rupture needs to be repaired or not. A thorough review of the literature (23) gave an amazing result: The percentage of excellent and good results tend to be in the same area, regardless of the therapy modality that was chosen. Therefore, there has to be a basic misconception, because a meaningfull operation should always do better than "doing nothing". Reasons for that may be based on the surgical technique, the sutures used or on the after treatment.
In this paper, we are focussing our interest on the suture material.

In order to learn something about the clinical relevance of resorption kinetics in the surgery of the knee, i.e. about the clinical use of the sutures investigated here, a questionaire was sent the Chiefs of Department of 248 Surgical Hospitals all over Germany: One of these questions inquired about the use of resorbable sutures in reconstructive surgery of the anterior cruciate ligament (ACL). Interestingly enough, 89.3% of all surgeons use that kind of sutures, among these PDS is the most common material (see Fig. 2). Not-resorbable synthetic sutures and especially steel - with the exception of cases dealing with children's osseous ACL-avulsions - no longer play a significant part in the operating theater. They all rely on the data for resorbable material given in the literature, e.g. a 50% reduction of maximum breaking strength in PDS within 5 weeks.

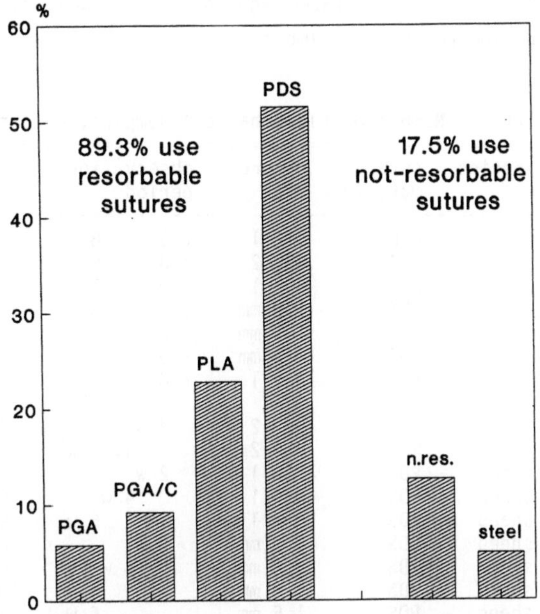

Fig. 2: Results of an inquiry of 248 Surgical departments: Predominantly resorbable suture materials are selected for ACL-repair. Sum (%) over "100" because of multiple quotations.

OBJECTIVES OF THE STUDY

The hypotheses to be tested were that resorbable suture materials show different kinetics of resorption depending on

- different implantation sites and
- different species.

MATERIALS AND METHODS

This paper comprises several pieces of work that have been done between 1985 and 1991. All of the animal experiments were performed according to the regulations of the Federal Law for Protection of the Animals and after Government allowance. Every painfull procedure was performed under general anesthesia, analgetics were administered as required.

The experimental groups and the numbers of samples for biomechanical testing are tabulated against suture material, suture size, observation period and site of implantation in Tab. 2.

Tab. 2: Number of biomechanical samples, n-total = 361

species	suture material	suture size	observation period		site of implantation s.c.	i.m.	i.a.	i.p.
-	PHB	USP 1	t 0	8	-	-	-	-
-	PLA	USP 2	t 0	8	-	-	-	-
-	PDS	USP 1	t 0	7	-	-	-	-
-	PDS	1.0 mm	t 0	8	-	-	-	-
-	PDS	1.5 mm	t 0	7	-	-	-	-
-	PDS	2.0 mm	t 0	7	-	-	-	-
-	PGA/C	USP 1	t 0	7	-	-	-	-
sheep	PLA	USP 2	2 w		8	8	8	-
sheep	PLA	USP 2	4, 6 w		not tested (nt)			
sheep	PDS	USP 1	2 w		7	8	-	-
sheep	PDS	USP 1	4 w		8	7	-	-
sheep	PDS	USP 1	6 w		9	8	-	-
sheep	PDS	1.0 mm	2 w		8	-	-	-
sheep	PDS	1.0 mm	4 w		8	-	-	-
sheep	PDS	1.0 mm	6 w		10	-	-	-
sheep	PDS	1.5 mm	6 w		8	-	-	-
sheep	PDS	2.0 mm	2 w		-	-	8	-
sheep	PDS	2.0 mm	4 w		8	-	16	-
sheep	PDS	2.0 mm	6 w		-	-	19	-
sheep	PGA/C	USP 1	2 w		8	9	8	-
sheep	PGA/C	USP 1	4 w		7	8	8	-
sheep	PGA/C	USP 1	6 w		5	7	nt	-
rat	PHB	USP 1	2 w		-	7	-	7
rat	PHB	USP 1	4 w		-	7	-	7
rat	PHB	USP 1	8 w		-	7	-	7
rat	PHB	USP 1	12 w		-	7	-	7
rat	PDS	USP 1	2 w		7	-	-	-
rat	PDS	USP 1	4 w		7	-	-	-
rat	PDS	USP 1	6 w		3	-	-	-
rat	PGA/C	USP 1	2 w		8	-	-	-
rat	PGA/C	USP 1	4 w		7	-	-	-
rat	PGA/C	USP 1	6 w		5	-	-	-
total / implantation site				**52**	**131**	**83**	**67**	**28**

Sheep: Under general anaesthesia the following implantations were performed in 8 female adult Merino sheep: 1) subcutaneous - in the region of the M. rectus lateralis; 2) intramuscular - M. quadriceps femoris; 3) intraarticular - medial to the femoral condyle (see Tab. 2).

Minipig: Under general anaesthesia the following implantation was performed in one minipig: subcutaneous - paramedian to the left M. rectus

abdominis, intramuscular - M. obliquus abdominis.
Rat: Under general anaesthesia the following implantations were performed in 49 female adult Wistar rats: subcutaneous (LÜNSTEDT and THIEDE) - in the region of the M. rectus abdominis and intraperitoneal (see Tab. 2).

Follow-up and material testing: After observation periods of 2, 4, 6 and 8 to 12 weeks (PHB the latter two), the animals were sacrified with an lethal overdose of pentobarbital, underwent gross pathological examinations and photographic documentation. The extremely careful harvesting of the implants, taking care not to damage the implants, was followed by biomechanical testing within two hours of sacrifice. The specimens (n=361) were

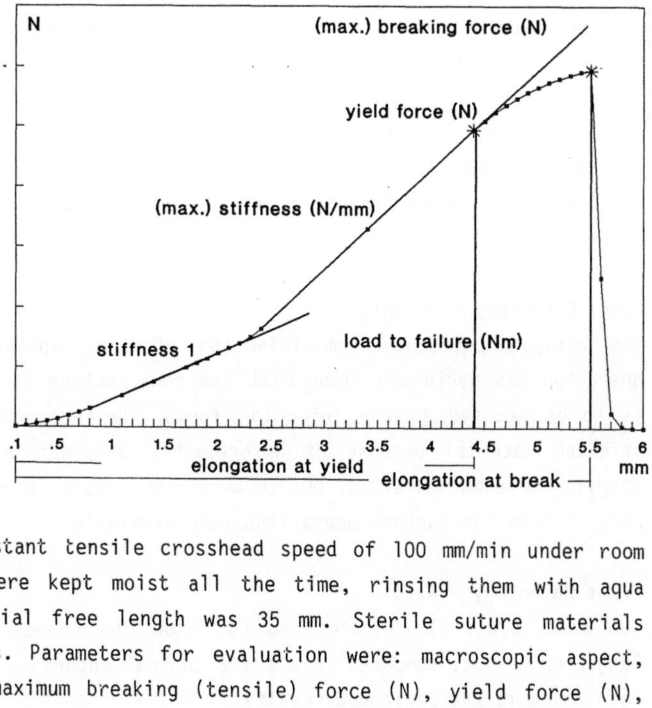

Fig. 3: Definition of biomechanical parameters. Load to failure is the area under curve, the integral calculus from the starting point to the perpendicular on the x-axis at the maximum breaking force

tested with a constant tensile crosshead speed of 100 mm/min under room temperature and were kept moist all the time, rinsing them with aqua bidestillata. Initial free length was 35 mm. Sterile suture materials served as controls. Parameters for evaluation were: macroscopic aspect, tissue reaction, maximum breaking (tensile) force (N), yield force (N), stiffness (N/mm), load to failure (Nm), mode of failure and elongation at break (mm) (Fig. 3). Where it seemed to be necessary, the results are given as normalized figures in percent of the respective controls. Several pilot experiments were run in order to estimate the systematical mistake that was encountered with our test set-up: Single suture loops with one surgical and five simple knots were tested with 8 initial length measurements (57 to 191 mm) and 5 tensile crosshead speed values (10 to 1000 mm/min.).

RESULTS

Gross Pathology - Sheep

All the sheep showed a small volume intraarticular effusion at sacrifice. There were no complications related to infection or to the operative procedure itself. All knee joints were free from arthrotic deformities.

Two weeks p.op. all implants were sheated with loose connective tissue, harvesting of the implants was done without any difficulties.

Four weeks p.op. the PLA-sutures reached a state of resorption that rendered it impossible to do any biomechanical testing. The PDS and PGA/C material was covered by more or less tight connective tissue, the removal of PGA/C from the joint cavity resulted in two material fractures.

Six weeks p.op. only dilusive traces of PLA were found. The tissue reaction towards PDS and PGA/C was less pronounced than at four weeks p.op.. No PGA/C material suited for biomechanical testing could be delivered from the knee joint.

Gross Pathology - Minipig

The minipig suffered from infection of the implantation sites. The operation was performed along with the construction of a gastrostomy that is to be accused as the infection focus. We resigned from testing the infected material because of unforeseable alterations in comparison to sterile implantation sites. Due to Government restrictions unfortunately no more animals for further operations were available.

Gross Pathology - Rats

Two weeks p.op. the sutures did not show any change of shape, colour, length or volume, compared to sterile control material. Moderate adhesions were found in the peritoneal cavity.

Four weeks p.op. a slight increase of the amount and texture of the connective tissue around the sutures was demonstrated.

Six weeks p.op. some sutures seemed to be incorporated in a taut scar, in the majority of cases not much of a difference compared to four weeks p.op. was found. The stains of the sutures persisted more in the PGA/C than in the PDS. Only three and five samples of PDS and PGA/C respectively were available for testing: The animals obviously succeeded in removing the material from the subcutaneous implantation bed.

Eight and twelve weeks p.op. the PHB had lost its staining. The adhesions were asymptomatic and consisted mainly of loose tissue of the omentum testis.

Biomechanics

Controls: The control values at t 0 are in a range of 75 to 86 % of what is given in the Ethicon Company's Product Information (6). In order to account for different test set-ups and methodological variances, the control values obtained by our group in the following are defined as internal standard, that is 100 % to normalize the p.op.-values.

PHB: PHB has very slow kinetics of resorption and demonstrates unfavourable biomechanical properties for application as a suture material: The force-lengthening characteristics are very poor in respect to knot tying safety and reliability of stress retaining. During the whole observation period the resorption in the peritoneal cavity is ahead of the intramuscular site.

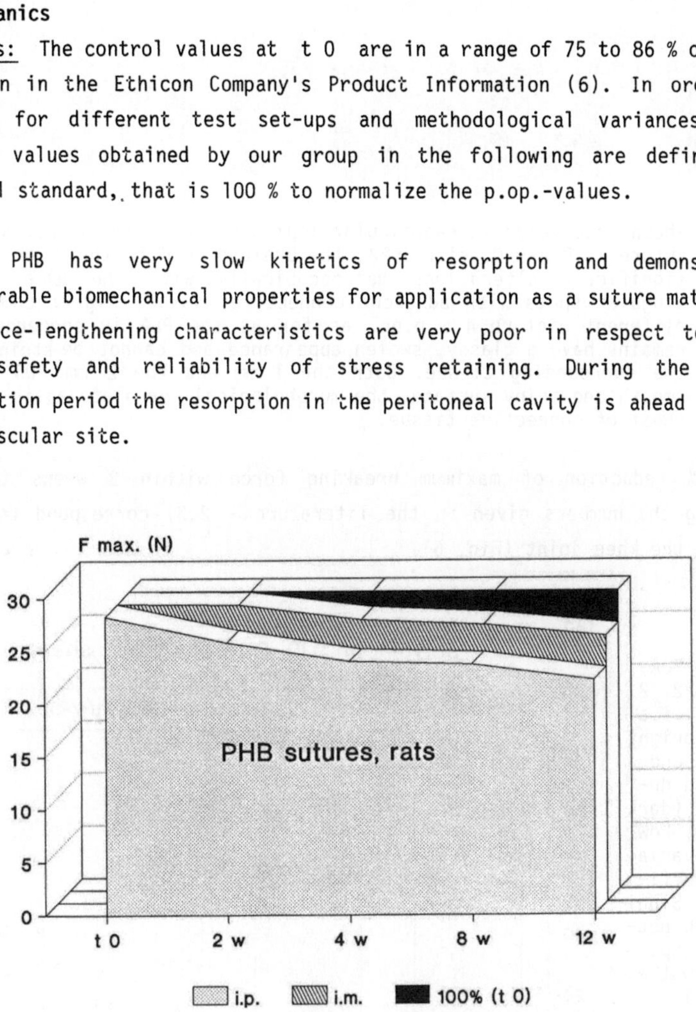

Fig. 4: PHB sutures, rat. I.m. vs. i.p. Implantation. Slow resorption, maximum breaking force 12 weeks p.op. is between 81% (i.p.) and 94.3% (i.m.). Faster degradation within the peritoneal cavity.

PLA: PLA-910 shows a clearcut difference between subcutaneous and intramuscular or intraarticular implantation. This is true for all biomechanical parameters two weeks p.op.. Four and six weeks p.op. the harvesting of this material for testing was impossible (Fig. 5).

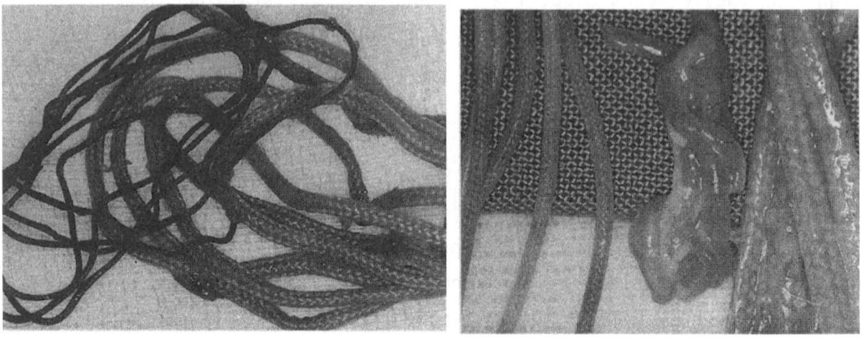

Fig. 5: Sheep, recovered intraarticular implants. Left: 2 w p.op., the dark suture is PLA-910, size USP2, the braided cord is the 2 mm PDS. No significant alterations macroscopically with the exception of singular spots with adherent connective tissue. Right: completely different picture 4 w p.op. as far as the PLA is concerned. The remains have a glassy, swolen appearance and cannot be stripped off the surrounding tissue. Both the 1 mm and the 2 mm PDS show a smooth and shiny surface, the morphological correlate again being a sheat of connective tissue.

The 48% reduction of maximum breaking force within 2 weeks (roughly matching the numbers given in the literature - 2,3) correspond to a 66% loss in the knee joint (Fig. 6).

Fig. 6: PLA, size USP2, 2 w p.op., i.a. means (bright shaded) and standard deviation (dark shaded). Low SD and variation coefficients. Significance between groups is below the 1% level.

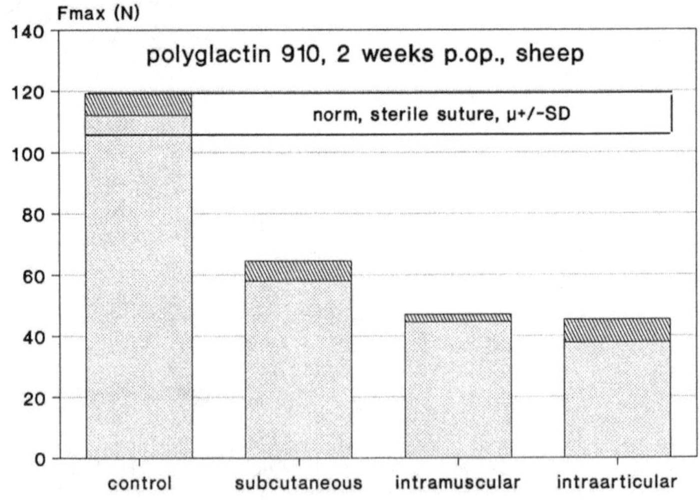

Not every biomechanical parameter is affected to the same extent. The load to failure is reduced to a final 16%, compared to 34% of breaking force and 70% of stiffness (Fig. 7). As the load to failure is an indirect measure of how much stress a suture repair can take in daily life, it can be concluded that such a repair is prone to fail within the articular cavity as early as 2 weeks p.op.

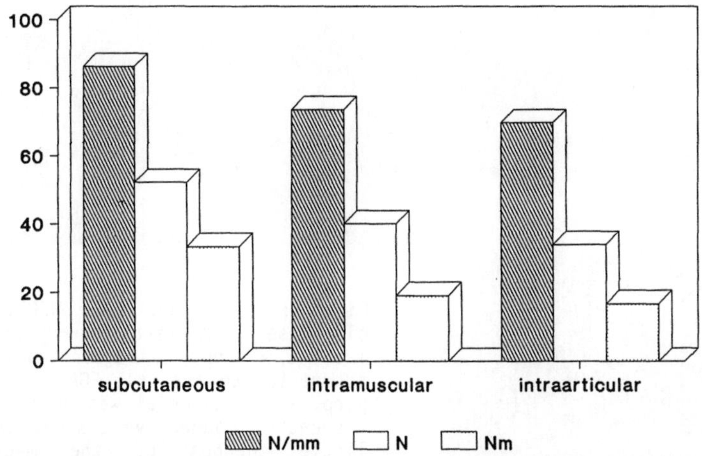

Fig. 7: PLA, USP2, sheep, 2 weeks p.op.; comparison (%) between the parameters breaking force, stiffness and load to failure of subcutaneous, intramuscular and intraarticular implants.

PGA/C: The breaking force of PGA/C at time 0 is higher in terms of absolute figures than that of PDS. The effective biomechanical halflife is definitively shorter than 21 days. 6 weeks p.op. the sutures became extremely brittle. On inspection, the sutures were covered by a thin layer of connective tissue, unaltered in position and configuration (Fig. 8). Removal from the intraarticular implantation bed however, that already 4 w p.op. had resulted in two material fractures, gave nothing but fragmented material, that broke under the conditions of blunt dissection. In this case the macroscopic aspect is by no means related to the poor material properties 6 weeks p.op.. Intramuscular and subcutaneous implants wound up with a breaking force of 12.7% to 19.1% and a residual stiffness of 44.4% (i.m., sheep), 50% (s.c., sheep) to 65% (s.c., rat) of normal controls.
The changes in the kinetics of resorption became more and more evident the longer the observation period was: 2 weeks p.op. no significant difference between the sites of implantation was demonstrated, whereas 4 and 6 weeks

p.op. the intraarticular "curve" diverged constantly from the other two (see Fig. 9 and 10).

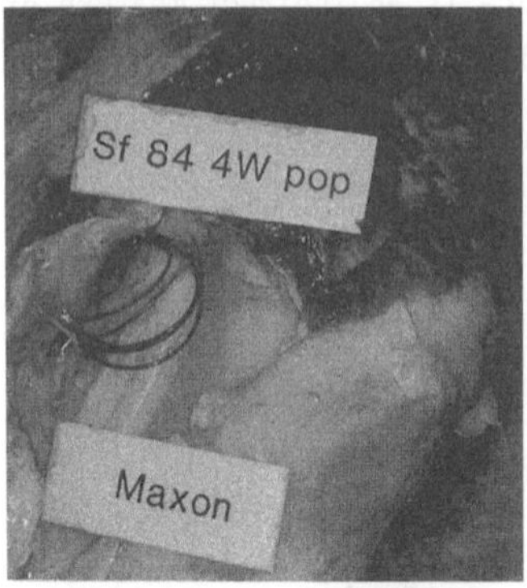

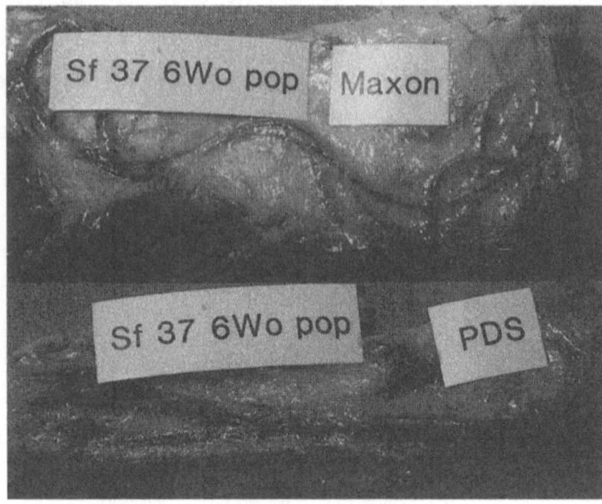

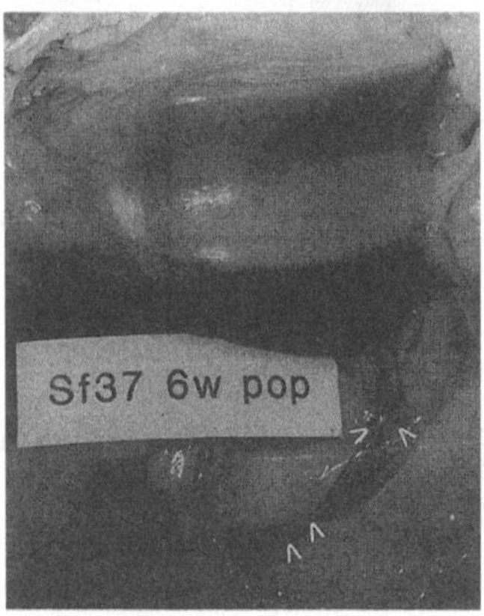

Fig. 8: PGA/C and PDS, USP1, sheep; clockwise from left to right: PGA/C, i.a., 6 w p.op. - fragmentation of the suture (white arrows); PGA/C, i.a., 4 w p.op. - the material was developed from a sheat of connective tissue, only very little changes of the macroscopic aspect; PGA/C and PDS, s.c., 6 w p.op. - in the subcutaneous area the scar looks more tight than in the knee joint; no adverse reactions of the host tissue.

Fig. 9: PGA/C sheep, max. breaking force (N). 2 to 6 weeks p.op. Little difference between curves 2 w p. op., increasingly accelerated resorption from the 4th week on. No samples could be tested 6 w p.op.

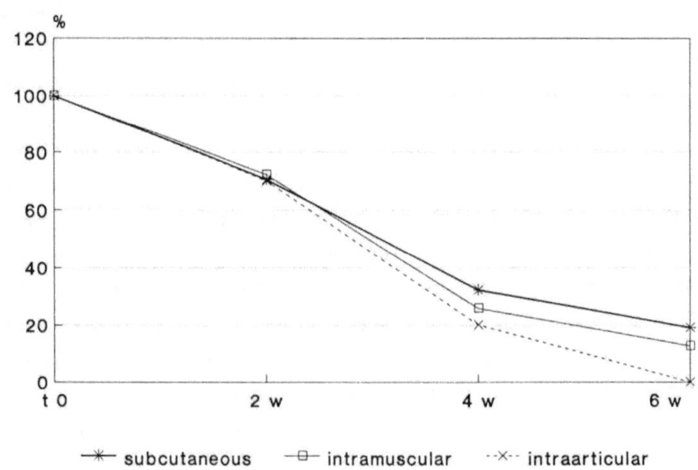

Fig. 10: PGA/C sheep, stiffness (N/mm). 2 to 6 w p.op. Like PLA, the parameter N/mm is rather stable - compared to breaking force. Significant loss of material properties do not occur before the 6th p.op. week. No test was performed of the 6 week i.a. samples.

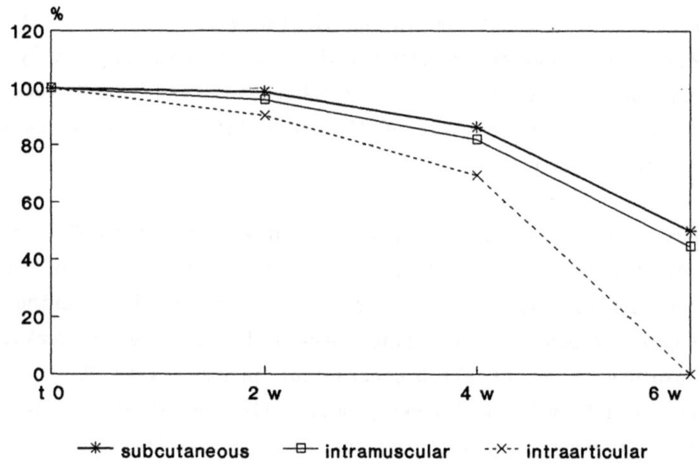

PDS and PGA/C both react in an analogous manner: Intraarticular implantation always leads to an accelerated deterioration of material properties.

<u>PGA/C - interspecies comparison:</u> Following subcutaneous implantation of size USP1 PGA/C, the course of the breaking force over time is retarded in the rat. 4 weeks p.op. this gap is the widest, 6 weeks p.op. implantation in both the rat and the sheep ends up with a 82% reduction of this parameter (Fig. 11). Changes of stiffness are less marked, but they show the same tendency - a slower resorption process in the rat.

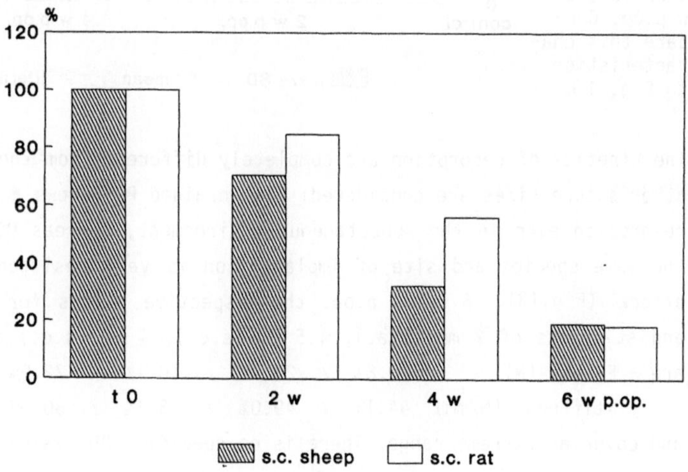

Fig. 11: PGA/C interspecies comparison rat vs. sheep, s.c. size USP1. Retarded resorption in the rat during the first 4 p.op. weeks. Same final value 6 w p.op. Breaking force in % of control

What was said about the intraarticular use of PLA also is valid for the PGA/C: The decline of biomechanical properties to roughly 45% within 21 days and the total loss of any inherent strength by 6 weeks p.op. cast a serious doubt on the use of this material for intraarticular ligament or meniscus repair.

PDS: The intraarticular resorption of 2 mm braided PDS is accelerated when compared to either the figures from the literature or to the data from the subcutaneous implants (Fig. 12). A 50% decrease in breaking force actually happens around the 3rd p.op. week and not after 5 weeks. Relative and absolute values of all biomechanical parameters of PDS are equal to those obtained from PGA/C 2 weeks p.op., while the latter material is stronger at time 0. From the 4th week on, PDS always has superior biomechanical qualities.

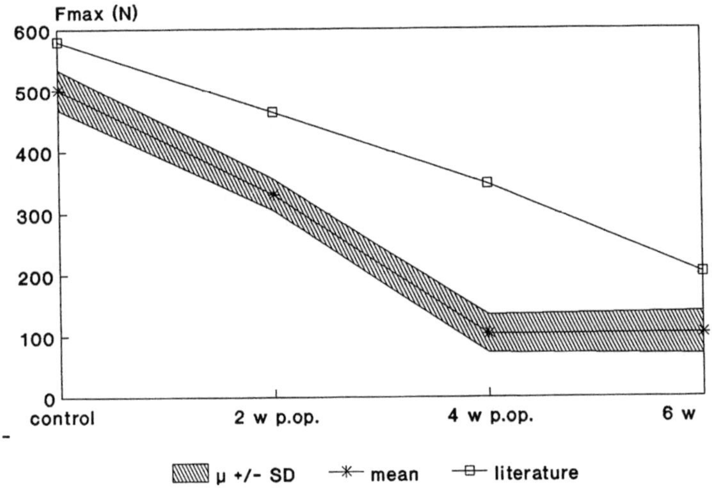

Fig. 12: PDS 2 mm, sheep, i.a. Fast resorption untill 4 w p.op. No further loss 6 w p.op. Compare this characteristics to Fig. 13.

The kinetics of resorption are completely different from the 2 mm PDS when other suture sizes are considered: 1 mm braided PDS shows a steady and fast resorption even in the subcutaneous environment, whereas USP1 material in the same species and site of implantation is very resistent towards host attack (Fig.13): 6 weeks p.op. the respective values for breaking force and stiffness of 2 mm (i.a.), 1.5 mm (s.c.), 1 mm (s.c.) and USP1 (s.c.) are - F max. (N) 20.8% / 23.0% / 10.1% / 72.9% and
- stiffness (N/mm) 44.1% / 49.0% / 15.1% / 50.0%

and cover an extreme range. There is no specific "PDS-value" to rely on.

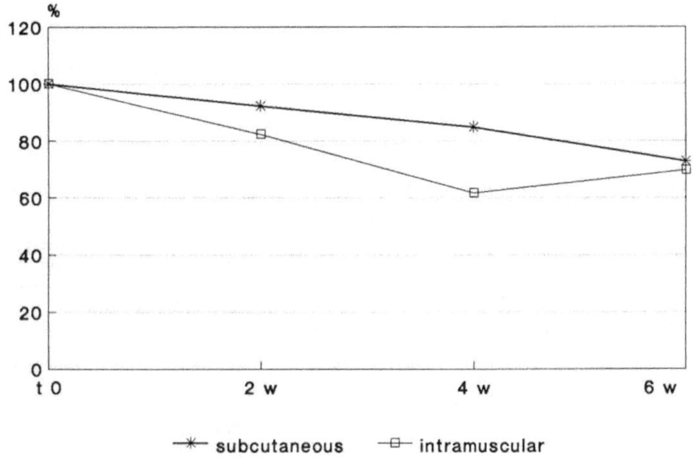

Fig. 13: PDS USP1, sheep, s.c. vs. i.m. very slow resorption compared to other size PDS material. Additional retardation s.c. vs. i.m. during the first 4 weeks p.op.

PDS - interspecies comparison: In contrast to the results of PGA/C there is no (clinical) significant difference between implantations into sheep and rat if only the size USP1 data are considered (insert Fig. 14). Reasoning that this looks completely different in case of e.g. 1 mm braided PDS is inadmissable because of the magnitude of the data variation (see above). Fig. 14 summarizes the interspecies comparison with PGA/C and PDS.

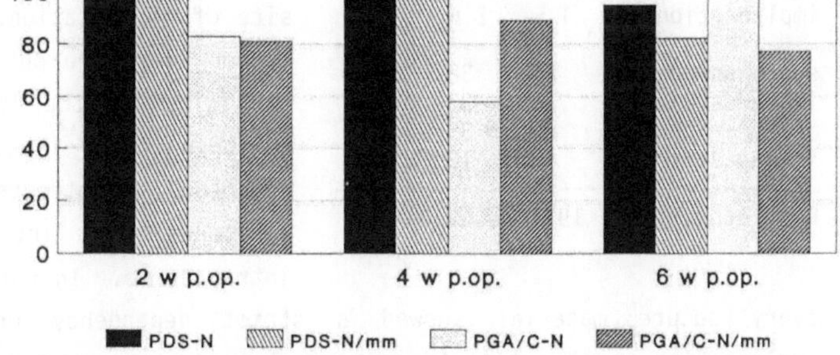

Fig. 14 - Insert: PDS, USP1, s.c., sheep vs. rat, 2 to 6 weeks p.op. Very similar kinetics during the observation period.

Fig. 14: Summary interspecies comparison sheep vs. rat, PGA/C and PDS, USP 1, breaking force (N) and stiffness (N/mm). Relative numbers calculated from the quotient sheep:rat over ´100´ are equivalent to retarded kinetics of resorption in sheep, numbers below ´100´ mean faster resorption (compared to rats). The correlation between biomechanics and species seem to depend on the suture material and the suture size.

DISCUSSION

Kinetics of Resorption Depending on the Site of Implantation

Contradictions in experimental literature on the use of resorbable sutures or suture augmentations (5, 7, 10, 11, 15-19) may be deducted on model specific differences of the resorption kinetics. It was proven that the kinetics of resorption of PLA-910, PDS and PGA/C - like with the resorbable drug carriers - depend on the site of implantation. This seems to be logical from a pathophysiological point of view: Though the resorption of these materials is based on hydrolysis - for the most -, the aggressive milieu of the joint cavity (1) along with an increased turn-over of the synovial fluid can be expected to interact more intensively with the material than in the subcutaneous area.

Tab. 3: Factors that (may) affect resorption/ degradation of aliphatic polyesters

chemical structure
configurational structure
molecular weight distribution
presence of low molecular weight compounds
shape
sterilizing process
morphology
annealing
storage history
site of implantation
species
adsorbed and absorbed compounds in vivo
physico-chemical factors
hydrolysis
stresses

These findings are in line with VERT´s summary of factors, that may affect the resorption of aliphatic polyesters (Tab. 1), the "species" should be supplemented to this table (Tab.3).

In 1975 POSTLETHWAIT demonstrated a correlation between the deterioration of material properties of PLA and the site of implantation. If the implantation sites are put in a line from the "slow" to the "fast" sites, the following order is established: Subcutaneous / intramuscular / intraperitoneal / intraarticular =/= intragastric. In this series every every suture material showed a strict dependency on the site of implantation or - certain biomechanical parameters - at least an apparent tendency towards that behaviour.

Tab. 4: Resorption kinetics depending on the site of implantation

site of implantation	F max. (%) / time 1 w	2 w	3 w
subcutaneous	80	38	10
intramuscular	75	33	9
intraperitoneal	63	35	12
intragastric	43	18	2

Postlethwait RW, 1975 (c.n. 3)

Kinetics of Resorption Depending on the Species

A correlation between the resorption and the species was established for

Tab. 5: Resorption kinetics of PDS depending on the species, 2 w p.op., s.c.

author	species	F max. (%)
2 Barrows	pig	77
	mouse	74
3 Chu	rat	74
13 Metz	rabbit	98
own results	sheep	65 - 92
	rat	86

PGA/C only. The experimental design did not include PLA and PHB into the interspecies comparison.
Nevertheless, the species will have to be taken into consideration when we talk about factors that influence- maybe in a negative manner - the outcome of operations with resorbable suture materials. Another argument in favour of this hypothesis is the review of the literatur (Tab. 5). The breaking force exemplifies the wide range of data for different species: 2 weeks post subcutaneous implantation values between 65% and 98% of the control were found by various investigators.

Implantation Technique

A "bulk" implantation (Fig. 15) like in this investigation (intraarticular) cannot be termed a very standardized method. Nevertheless, it had to be done that way because it was considered unethical und uneconomic to sacrifice one sheep per 8 cm suture-sample. The localisation dorsomedial to the femuropatellar joint seems to be a fairly good choice, none of the implants demonstrated mechanical lesions that could be attributed to impingement. The strain on these materials in the clinical situation of an augmented ACL-repair for example is probably x-fold: Through every range of motion the implant suffers tensile, rotational (40° to 120°) and shear stress, the orifices of the drill canals through the bone are sharp stress risers. Therefore one should bear in mind, that the results of the intraarticular site of implantation may be even false positive.

Biomechanical test methods

Criticism towards the biomechanical testing used in this study seems justified when considering that the majority of samples failed at or near fixation devices. According to DIN 53857 all of these tests would be invalid, for samples that fail at or within 5 mm of fixation would have to be rejected. On the other hand, the first and foremost intention in this series was not to obtain absolute values, but to estimate - in relative terms - the kinetics during a certain period of time. The systematical mistake is the same in all tests; the systematical mistake can be assessed

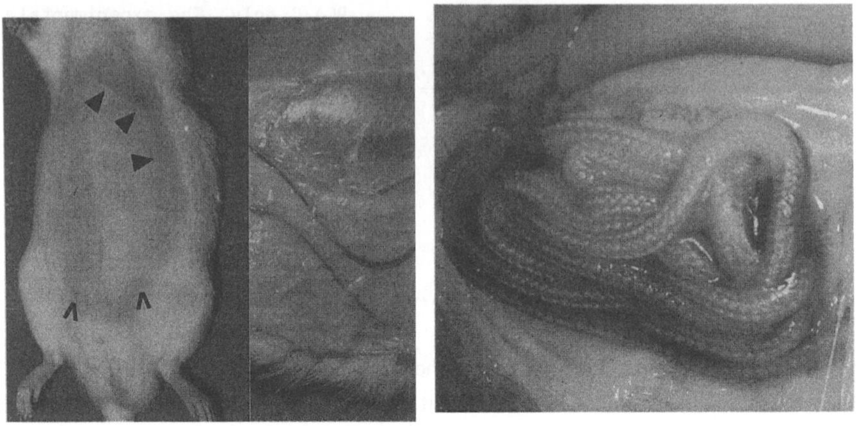

Fig. 15: Left: The best and most standardized method of implantation of sutures was described by LÜNSTEDT (12) and others. At sacrifice, the suture - that was implanted with a minimal tissue trauma - lies straight and isolated in the host bed. Richt: In "bulk" implantations mechanical and biochemical interference of the material with itself is possible.

by comparing our values to other biomechanical tests methods. If samples fail at the fixation, the results are somewhat lower than in midsubstance failure and equal the knot holding capacity. This critical mode of failure was not only accepted but even favoured: It enabled us to know the initial free length of the test material, which is unknown when "wrap around"-techniques are used (14). The initial free length is a prerequisite to compare the stiffness of different samples, because it is the most important single factor to interfere with the parameter "stiffness" (Fig. 16).

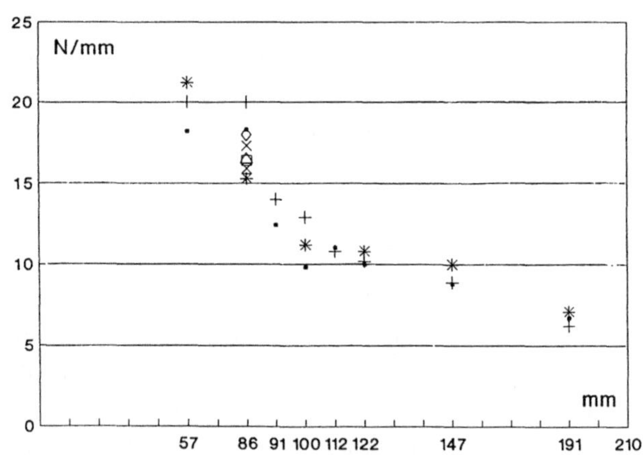

Fig. 16: Stiffness of USP1 sutures vs. initial free length, n = 23, crosshead speed 100 mm/min.
The correlation between stiffness and initial free length is more or less trivial, but it has to be established for each and every test system.

The variation coefficient usually was around 6 to 9 % which we consider to be a good result for a biological system.

Review of the Literature

Unlike the PLA (22), there seems to be an inverse correlation between resorption and suture size with the PDS material - if any at all. Both PDS and PGA/C show a non-linear behaviour; we have no explanation for the increase of stiffness over normal two weeks p.op. (PGA/C).

VASENIUS et al (21) investigated SR-PGA composite rods for fracture fixation. They saw a completely different behaviour of the reduction of bending strength over time when the tests were performed in vitro and in vivo. Coating with various PLA-composites and PDS did not automatically result in markedly retarded resorption of PDS-coated SR-PGA-rods vs. PLA-coatings.

Tab. 6: Review of the literature on kinetics of resorption

author		material	F max. (% of control) / t (w p.op.) 2	3	4	5	6	7
8	Gogolewski	Polydioxanon	-	60	-	40	-	-
2	Barrows	Polydioxanon	74	-	-	-	-	-
13	Metz	Polydioxanon	98	110	38	84	-	-
3	Chu (Ray)	Polydioxanon	74	-	58	-	41	-
15/6	Rehm	Polydioxanon	-	-	-	-	58	-
own results		**s.c.**	**65-92**	**-**	**31-79**	**-**	**10-77**	**-**
		i.m.	**82**	**-**	**62**	**-**	**70**	
		i.a.	**65**	**-**	**21**	**-**	**21**	**-**
8	Gogolewksi	Polyglycol-Co-Tri-methylencarbonat	-	-	55	-	-	14
13	Metz	Polyglycol-Co-Tri-methylencarbonat	88	45	23	11	-	-
own results		**s.c.**	**71-85**	**-**	**32-56**	**-**	**18-19**	**-**
		i.m.	**72**	**-**	**26**	**-**	**13**	**-**
		i.a.	**70**	**-**	**20**	**-**	**0**	**-**
8	Gogolewski	Polyglycol	30	-	-	-	-	-
8	Gogolewski	Polyglactin	-	30	-	-	-	-
2	Barrows	Polyglactin	55	-	-	-	-	-
3	Chu (zit.)	Polyglactin	38	10	-	-	-	-
20	Thiede	Polyglactin	34	20	6	-	-	-
own results		**s.c.**	**52**	**-**	**0**	**-**	**0**	**-**
		i.m.	**40**	**-**	**0**	**-**	**0**	**-**
		i.a.	**34**	**-**	**0**	**-**	**0**	**-**

Our results with PDS are well within the values quoted in the literature, but the intraarticular numbers are the lowest, the s.c. implants in the sheep the highest of this survey. PGA/C characteristics are the same like in the literature, again with the exception of the intraarticular site. PLA also equals literature data, i.a. results at the lower limit.

Conclusions

The crucial question in experimental surgery is: To what extent and under which conditions can experimental results be transferred to the human counterpart?. The proven dependency on the site of implantation and -sometimes- on the species implies, that the e.g. "50 % reduction of breaking force in five weeks" for PDS (15, 16) is no longer a safe and readily acceptable statement. No one knows for sure, what is more equivalent to a man's knee: The subcutis of the rat or the knee of the sheep. Nevertheless, to be on the safe side, from our experiments, we conclude that the intraarticular use of PLA and PGA/C is obsolet. 3 weeks p.op. these suture materials have only neglegible biomechanical properties (breaking force, load to failure) left. Neither ligament nor meniscus heal within 3 weeks to obtain inherent strength as to tolerate even daily life´s strains.

The only resorbable suture material that one should rely on within the articular cavity, is polydioxanone. Even so, in the light of early postoperative motion, a 50 % reduction in three weeks is a solid basis to count on and select the size of your sutures.

The intraarticular resorption characteristics, to our best knowledge not investigated until today and unknown to every surgeon, may constitute one of the more important reasons for eventual failure of intraarticular ligament and meniscus surgery.

LITERATURE

1 Ascherl R, Tschesche H, Staudhammer A, Wendt P, Lechner F, Blümel G (1986) Die Neosynovialflüssigkeit nach alloplastischem Gelenkersatz und Implantatlockerung, H z Unfallheilkunde 181:194-198

2 Barrows Th, Grussing DM, Havens MA (1985) Comparison of tensile and knotted strength retention of synthetic absorbable sutures in vivo. In: Ratner BD, Soc Biomat (eds.): Transactions Soc Biomat Vol. VIII: 143

3 Chu CC (1985) The degradation and biocompatibility of suture materials. In: CRC Critical reviews in biocompatibility, CRC Press Inc, Boca Raton, Florida USA, Vol 1 No 3: 261-322

4 Deutsche Industrie Norm DIN 53857 (1979) Beuth Verlag Berlin Köln, 09.79, DK 677.074- 418: 620.172

5 Diehl K, El-Ahmad M, Franzl K (1987) Kapselbandchirurgie des Kniegelenkes mit resorbierbaren Materialien. 5 Jahre klinische Erfahrung mit Vicryl- und PDS-Bändern, Z Orthop 125: 467-472

6 Ethicon Company Product Information (1990), 58Le78, 5Le3/88, 5Le1/89,2-7

7 Friedrich A, Wolter D (1985) Der resorbierbare alloplastische Bandersatz mit Polyglactin - Experimentelle Ergebnisse und klinische Bedeutung, Zbl Chirurgie 110: 908.8791

8 Gogolewski S (1990) Resorbable polymers for internal fixation. In: Hofmann GO (Hrsg.): "Praxis-Forum" Medizintechnik 20: 72-82

9 Hamelmann H, Thiede A (1982) Übersicht zur Wahl des Nahtmaterials. In:

Thiede A, Hamelmann H (Hrsg.) Moderne Nahtmaterialien und Nahttechniken in der Chirurgie, Springer Berlin Heidelberg New York, 418-422

10 Hoffmann R, Lobenhoffer P, Krettek C, Tscherne H (1989) PDS (Polydioxanon)- Augmentation der vorderen Kreuzbandrekonstruktion - Eine experimentelle Stabilitätsuntersuchung, H z Unfallheilkunde, Springer Berlin Heidelberg, 207: 274

11 Holzmüller W, Rehm KE, Perren SM, Rahn B (1989) Das PDS-augmentierte Patellarsehnentransplantat zur Rekonstruktion des vorderen Kreuzbandes am Schafsknie, Langenbecks Arch Suppl Chir Forum, Springer Berlin Heidelberg, 265-268

12 Lünstedt B, Thiede A (1982) Standardisierung der Nachweisverfahren zur Objektivierung der linearen Fadenzug-, Knotenbruch- und Knotensitzfestigkeit verschiedener absorbierbarer und nichtabsorbierbarer Nahtmaterialien. In: Thiede A, Hamelmann H (Hrsg.): Moderne Nahtmaterialien und Nahttechniken in der Chirurgie, Springer Berlin Heidelberg New York, 16-25

13 Metz SA, Chegini N, Masterson BJ (1990) In vivo and in vitro degradation of monofilament absorbable sutures PDS^R and Maxon^R. Biomaterials 11: 41-45

14 Metz SA, Fraunhofer JA v, Masterson BJ (1990) Stress relaxation of organic suture materials, Biomaterials 11:197-199

15 Rehm KE, Schultheis KH, Bopp P, Ecke H (1984) Biomechanische Untersuchungen vom resorbierbaren Bandersatz und deren klinische Bedeutung, Langenbecks Arch Suppl Chir Forum, Springer Berlin Heidelberg, 207-211

16 Rehm KE, Schultheis KH (1985) Bandersatz mit Polydioxanon (PDS^R). Unfallchirurgie 11:264-273

17 Schöttle H, Apel R, Kilgus O (1989) Die Sicherung der vorderen Kreuzbandnaht durch PDS-Kordel, H z Unfallheilkunde 207:439

18 Schöttle H, Meenen NM, Kilgus O (1990) Bandverstärkung mit resorbierbarer PDS-Kordel und frühfunktioneller Nachbehandlung. Unfallchirurg 93:35-39

19 Siebels W, Ascherl R, Schwerbrock R, Maurer M, Blümel G (1989) Die Auswirkung von temporären synthetischen Verstärkungsmaterialien auf die biomechanischen Eigenschaften gestielter Patellarsehnenplastiken als Kreuzbandersatz beim Schaf, Langenbecks Arch Suppl Chir Forum, Springer Berlin Heidelberg, 261-264

20 Thiede A (1982) Biologische Wertigkeit der Nahtmaterialien. In: Thiede A, Hamelmann H (hrsg.): Moderne Nahtmaterialien und Nahttechniken in der Chirurgie, Springer Berlin heidelberg New York, 238-252

21 Vasenius J, Vainionpää S, Vihtonen K, Mero M, Mikkola J, Pellinen M, Rokkanan P, Törmälä P (1991) Biodegradable self-reinforced polyglycolide (SR-PGA) composite rods for fracture fixation coated with slowly biodegradable polymers - strength and strength retention in vitro and in vivo. In: Barbosa MA (ed) Biomaterials degradation, Elsevier Science Publishers Amsterdam Oxford New York Tokio, 393-395

22 Vert M, Guerin Ph (1991) Biodegradable aliphatic polyesters of the poly (hydroxy acid)-type for temporary therapeutic applications. In: Barbosa MA (ed) Biomaterials degradation, Elsevier Science Publishers Amsterdam Oxford New York Tokio, 35-51

23 Wirth CJ (1989) Kreuzbandverletzungen des Kniegelenkes, Orthopäde 18:302-314

SELECTED DISCUSSION REMARKS:

Planck: We all know that there is no correlation between in vitro tests and the in vivo behaviour. Now you as a surgeon say there is no correlation between different implantation sites and different species. So in which way should we carry out the tests for a device?

Scherer: Try to standardize the test methods all over the world, that is the only way to get comparable results.

Gibbons: Do you have any explanation for the differences in degradation at different implantation sites?

Scherer: The most important factor for the differences in degradation at different implantation sites probably is the substance turn-over. If you implant something in the subcutaneous environment, you have a low vascularity over a certain period of time, you don't have much fluid exchange and you don't have many enzymes near the implant. On the other hand after an implantation in a knee joint, for example, there is a sharp increase of all sorts of enzymes from macrophages, lysosomal enzymes, lipases, phospholipases, and a very large turn-over of the joint fluid. If you put it into an order, then you have the subcutaneous, the intramuscular, the intraperitoneal and probably equivalent, that is at least what Postlethwait says, the intragastric and intraarticular environment.

TISSUE RESPONSE TO RESORBABLE SYNTHETIC POLYMERS

D. F. GIBBONS

BIOSCIENCES LABORATORY, 3M CENTER, ST. PAUL, MN. 55144

ABSTRACT:

When the soluble degradation products from currently available resorbable synthetic polymers are evaluated in vivo the chronic inflammatory response appears to be independent of chemical composition. The molecules such as cytokines, interleukins and growth factors which are secreted by the macrophage/giant cells during phagocytosis may well vary with composition. The biological responses observed so far appear acceptable and leave little or no local evidence of their presence after all of the material has departed.

Keywords: degradation, resorbable, synthetic polymer, inflammatory response, sutures, implantable devices.

INTRODUCTION:

In this paper we shall be addressing only the biological issues related to resorbable synthetic polymer systems. That is polymers which are capable of going into solution in vivo and being excreted in feces, urine or expired gases via the lungs either as monomers, oligomers or biotransformed products, Table 1. This is in contradistinction to degraded polymers whose molecular weight, and as a result mechanical properties, decrease so that fragmentation results but the particles remain in macrophages and foreign body giant cells in situ or regional lymph nodes and do not go into solution, examples include poly-ethylene teraphalate (Dacron R or Terylene R), nylons, polyaramides (Nomex, R) etc. Medical and Dental applications for resorbable polymers vary from wound closure, small vessel ligation, internal fracture fixation to drug delivery. The specific use of the resorbable polymer influences the mode of fabrication and, therefore, the polymer structure parameters such as, polymer chain orientation, crystalline/amorphous ratio. If strength is important, as in sutures or any load carrying fiber, then the degree of polymer chain orientation and degree of crystallinity will be high. The control of these parameters in injection molding which is used for fabrication of rods or plates, screws, etc. used in orthopedic applications depends upon the precise details of the molding operation as well as any thermal treatments used after molding. In applications where mechanical strength is of less critical importance, such as a vehicle for drug delivery, the device may be fabricated

H. Planck M. Dauner M. Renardy (Eds.)
Degradation Phenomena on Polymeric Biomaterials

TABLE 1

Polymer	Degradation Time/Uses
Poly D-L Lactic Acid	Weeks; amorphous
Poly L Lactic Acid	Months to years depending upon fabrication; plates, fibers
Poly glycolic acid	Weeks; sutures (Dexon R)
Poly lactic-co-glycolic acid	Weeks to months depending on ratio; sutures (Vicril R), staples and ligating clips.
Poly dioxanone	Months; (PDS R) sutures
Poly (glycolide-co-trimethylene carbonate)	Months; (Maxon R) sutures
Polyhydroxybutyrate-co-valerate	Years;
Polyester amide	Months to years; fibers
Poly (anhydrides)	Months; hydrophobic; drug delivery

by solution casting or precipitation from solution, where control of crystallinity and crystallite size may be more difficult to manipulate. In the case of hydrophilic resorbable materials water will swell predominately the amorphous phase where it will act as a plasticizer and in certain systems may allow sufficient molecular chain mobility to induce crystallization (1).

All of these polymer structure parameters are important in the control of the biological response since they control access and rate of access of water and, therefore, the hydrolytic degradation rate of the device. The biological response will be controlled by the surface properties of the device and the concentration of the soluble products of degradation in the surrounding tissue. For hydrophilic materials the rate of production of soluble degradation products is controlled by the solubility of water in each of the phases present. The solubility of water is greatest in the amorphous phase and neglible in the crystalline phase, hence the importance of controlling the phases and their distribution in any resorbable device.

Finally before discussion of the biological response to resorbable synthetic materials we must recognize the sequence of events which occurs during hydrolysis. As the molecular weight of the polymer decreases by hydrolysis the mechanical strength of the material decreases (2). However, there is no significant weight loss at this time since the molecular weight of the polymer chains is still too high to be soluble. Only when the molecular weight of the chain is decreased to very low values do they have significant solubility. A number of recent studies (2) (3) (4) (5) suggest no significant weight loss until the average molecular weight is below ~5 kD and much less than that for significant solubility, the actual values depend upon the chemical composition of the monomer. At the beginning of observable weight loss we can anticipate that small particles break away from the main body of the device since the mechanical strength is markedly decreased while the solubility is still marginal.

BIOLOGICAL RESPONSE:

The early literature on the biological response to synthetic resorbable polymers originated from their use as sutures. The work of Salthouse and others (6) demonstrated minimal inflammatory response to the resorbing suture. The evaluation of the biological response to individual fibers in subcutaneous sites represents the basis for these data. Figure 1(a) illustrates the response to a non-resorbable fiber, polypropylene and Figure 1(b) that from a poly L-lactic fiber of similar dimensions. In each case a macrophage or foreign body giant cell is in immediate contact with the fiber surface and a thin collagenous fibrous capsule surrounds the fiber. Before significant degradation and weight loss occurs it is difficult to differentiate between noncytotoxic resorbable materials by means of tissue response. If a three dimensional porous structure is fabricated from fibers in the 10 to 40 micron diameter size range it is possible to follow the tissue response from the initial

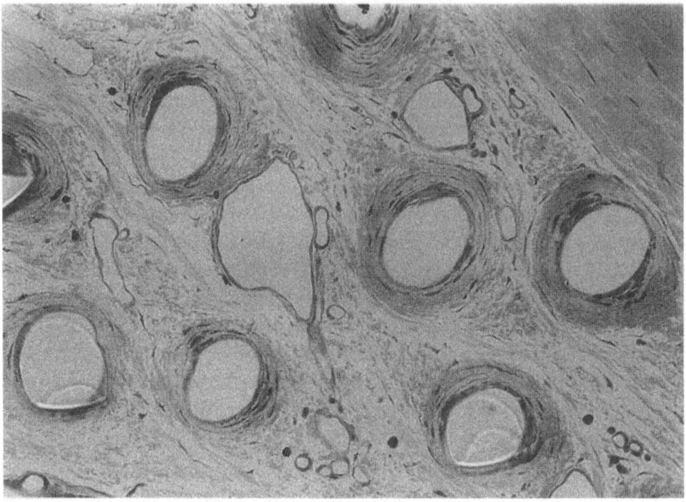

FIGURE 1(a) TISSUE RESPONSE TO POLYPROPYLENE FIBERS(ORIG. MAG. x 84)

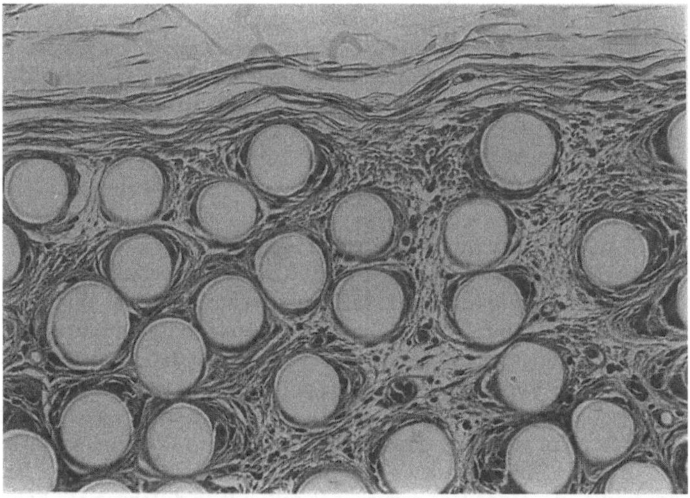

FIGURE 1(b) TISSUE RESPONSE TO POLY L-LACTIC ACID FIBERS AFTER 6 MONTHS (ORIG. MAG. x 84)

tissue ingrowth through complete disappearance (disolution) of this structure. This has been carried out for polyglycolic acid structures (7) and polylactides (8). The sequence involves the initial invasion of capillaries and fibroblasts and similar to the individual fibers, those in the structure are surrounded by multinucleated giant cells (MNGC's). As the material looses strength the structure collapses and small particles break off from the fibers and are phagocytised by macrophages and MNGC's. Light microscopy reveals these cells to have a typical foamy cytoplasm. Finally when all the material and particulate have resorbed and been excreted the invading capillaries atrophy, the macrophage/MNGC's disappear and only the areolar collagen from the interstices remains, eventually to be remodeled in accordance with the local site architecture. In the example described we are lead to the conclusion that since no necrosis or crytotoxicity was observed the concentration of soluble monomer and/or oligomer was not of sufficient concentration, even in the MNGC's, to reach the level which would induce these responses.

In order to ascertain whether this conclusion is more universally correct for two different resorbable systems, namely poly L-lactic acid and polyester amide we have hydrolysed fibers in vitro and collected the particulate material generated. Using such particulate materials, where the average molecular weight was reduced to less than 5.0 kD and the peak particle size was approx. 40 micron, they were implanted into a subcutaneous pocket in rats using a trochar, so that the pocket contained a concentrated bolus of particles. After one week the response Figure 2 was more intense but similar in structure to that observed in the three dimensional structure described earlier. In the periphery, where the particle density was less, the response was entirely similar to that from the three dimensional structure. Further work is underway to characterize the biological response using ultrastructural data from transmission-microscopy and histochemical techniques.

When more substantial devices are evaluated such as bone plates, etc. similar biological responses are observed (9). The only difference would appear to be the presence of a serous rather than cellular region at the interface in somecases(10).

DISCUSSION:

It must be noted that care should be taken when evaluating the biological response in mobile tissue such as the dermis or muscle to any fiber, including those which are resorbable. The presence of increased cellularity together with an increase in thickness of the fibrous capsule (collaring) may be the result of local mechanical trauma caused by relative motion between the fiber and local tissue and produced by a difference in mechanical compliance between fiber and tissue (11).

Although more data characterizing the biological response of macrophages and MNGC's while phagocytosing of resorbable particles is needed some

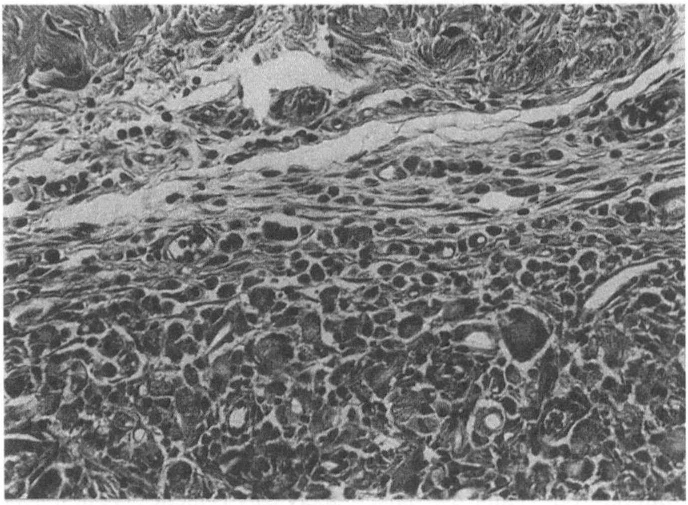

FIGURE 2(a) SUBCUTANEOUS TISSUE RESPONSE TO PLA PARTICLES < 40 MICRA, FORMED BY HYDROLYSIS OF FIBERS AT 80°C AND pH 11.00 AFTER 66 HOURS (ORIG. MAG. X 84)

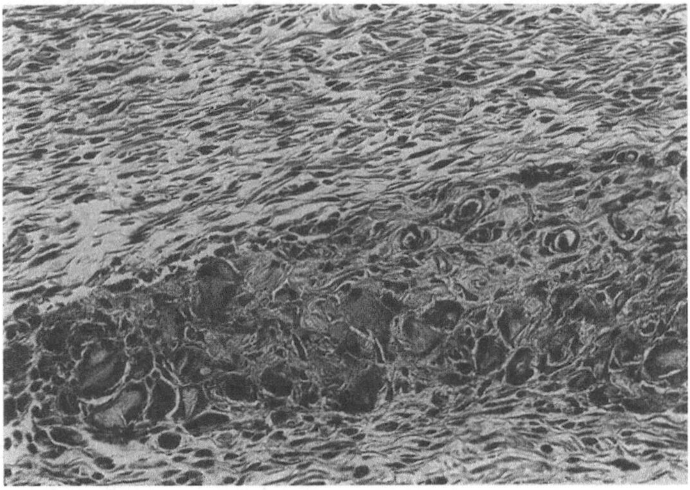

FIGURE 2(b) SUBCUTANEOUS TISSUE RESPONSE TO POLYESTERAMIDE PARTICLES < 40 MICRA, FORMED BY HYDROLYSIS OF FIBERS AT 80° C AND pH 11.00 AFTER 4 HOURS (ORIG. MAG. x 84)

initial data (12) (13) demonstrates that with polyglycolic acid particles they secrete growth factors such as TGF beta. The response of macrophages to textured surfaces has also demonstrated the secretion of growth factors with characteristics similar to PDGF (14). Care should be exercised, however, in extrapolation of these data since in vitro inhibitatory factors have been identified with particles from vascular grafts fabricated from polyethylene teraphalate containing delusterant particles of titania (12).

In general, therefore, the biological response to a number of resorbable synthetic polymers appears to be mild and result in no lasting deleterious effects when the volume of material undergoing rapid dissolution is relatively small. However, with larger volumes such as from internal fixation plates, screws, and knee ligament reconstruction devices, further evaluation work needs to be undertaken. Preliminary work on the evaluation of a resorbable Kennedy R LAD ligament augmentation device in sheep (15) and goats (16) suggests that the biological response is similar to that already described and therefore will not initiate an unacceptable response.

I wish to acknowledge T. H. Barrows and M. T. Truong, 3M Center for providing the particles of degraded poly L-lactic acid and poly esteramide; C. L. Van Kampen, 3M Center and Y. Kawamoto 3M/Sumitomo Japan for data on the poly L-lactic acid fibers used to evaluate the biological response. I wish also to thank G. S. Nelson, histo- technologist in our laboratory for her expertise in preparing the tissue sections for evaluation.

REFERENCES:

(1) Vert M, "Degradation of Polymeric Biomaterials with Respect to Temporary Therapeutic Applications: Tricks and Treats" (1990) Degradable Materials editors Barenberg SA, Brash JL, Narayan R and Redpath AE, CRC Press, Boca Raton, page 11

(2) Dijkstra PJ, Eenink MJD and Feijen J "Alpha-Hydroxy Acid Based Biodegradable Polymers Poly (L-Lactide) and Polydepsipeptides" (1990) Degradable Materials editors Barenberg SA, Brash JL, Narayan R and Redpath AE, CRC Press, Boca Raton, page 99

(3) Helder J, Dijkstra J and Feijen J. "In Vitro Degradation of Glycine/DL-Lactic Acid Copolymers", (1990) J. Biomed Mater Res 24:1105

(4) Sehakenraad JM, et al. "In Vivo and In Vitro Degradation of Glycine/DL-Lactic Acid Copolymers" (1989) J Biomed Mater Res 23:1271

(5) Leenslag JW, et al. "Resorbable Materials of Poly (L-Lactide), In Vivo and In Vitro Degradation" (1987) Biomaterials 8:311

(6) Salthouse TN, "Biologic Response to Sutures" (1980) Otolaryngol., Head, Neck Surg 88:658

(7) Gibbons DF and Lewis TW "Biodegradation of a Poly(Glycolic Acid) Epineural Tubulization Device" (1990) Degradable Materials editors Barenberg SA, Brash JL, Narayan R and Redpath AE, CRC Press, Boca Raton page 405

(8) Nakamura T, et al, "Bioabsorption of Polylactides with Different Molecular Properties", (1989) J Biomed Mater Res 23:1115
(9) Vert M et al (1984) "Macromolecular Materials", Hastings GW and Ducheyne P eds. CRC Press Boca Raton page 119
(10) Vert M, Private Communication
(11) Gibbons DF, Van Kampen CL, Mendenhall HV and Lamprecht EL "Effect of Mechanical Motion on the Tissue Response to Fibers" to be published
(12) Greisler HP et al "Macrophage/Biomaterial Interactions: The Stimulation of Endothelialization",(1989) J Vasc Surg 9:588
(13) Greisler HP, The Role of the Macrophages in Intimal Hyperplasia", (1989) J Vasc Surg 10:566
(14) Korman WJ, Gibbons DF and Sudilovsky O, "Effect of Surface Morphology on Exudate Composition and Biological Response", (1983) J Biomed Mater Res 18:225
(15) Christel P, Therin M, Mendenhall HV, Gibbons DF, "Tissue Response to Polylactic Acid Braided LAD in Sheep" to be published
(16) Hara N, Mendenhall HV, Kawamoto Y, Gibbons DF, and Van Kampen CL, "Tissue Response and Mechanical Properties of Polylactic Acid Braided LAD in Goats" to be published

SELECTED DISCUSSION REMARKS:

Stokes: In the example where you put a felt around the nerve, you presented the response of the tissue outside the nerve. But what happened to the nerve?

Gibbons: The nerve was completely unchanged.

Missirlis: Did you see some evidence that the mismatch of mechanical properties, especially the stiffness, is the stimulus for the proliferation of monocytes or other phagocytic cells around the polymer fibers?

Gibbons: I think that it is responsible for the stimulation of fibroblasts, not necessarily for an increase in monocytes.

Missirlis: Do you have an idea which mechanism is responsible for the proliferation of the cells?

Gibbons: I have two possible explanations:
One is, that the mechanical motion and proliferation alters slightly the genome of the fibroblast. The second explanation is the enhanced secretion of collagen and pro-collagen, that is all organized extra-cellularly. It is known, that low molecular weight collagen leads to an increased cellular response. In staining experiments we could show low molecular weight collagen at the implant site.

Kricheldorf: A potential degradation product of polyglycolic acid is glycolic acid, which has a reactive aldehyde group, which can react with surrounding protein and cause necrosis. Did you see any evidence for that degradation pathway and did you see an increasing amount of necrosis around polyglycolic acid?

Gibbons: No, I didn't see any necrosis in our experiments with the nerve cuffs made from polyglycolic acid.

IN VITRO DEGRADATION OF POLYLACTIDES DEPENDING ON DIFFERENT PROCESSES

M. Dauner, E. Müller, B. Wagner, H. Planck

Institut für Textil- und Verfahrenstechnik
Körschtalstr. 26, D-7306 Denkendorf

ABSTRACT:

Two Poly-L-Lactides from different manufacturers and of different molecular mass and one poly-D-L-Lactide polymer were processed to fibers, sheets and nonwoven. Following an accelerated 80°C-degradation test, they were evaluated regarding their degradation behaviour measuring the decrease of the molecular mass, the strength and the substance mass. While the dependance from the polymer composition and the molecular mass was evident, no significant effect from the crystallinity and the surface/volume relationship could be detected.

INTRODUCTION:

Besides a lot of other factors the degradation kinetics depends on the polymer composition and the molecular mass of the material, the morphology and the structure of the device, and the influence of thermal processes. The structure and design of the device determines the processing of the polymer and influences to a high degree its morphology, namely the crystallinity. And it defines the surface exposed to the environment in relation to the volume of the device.

MATERIAL AND METHODS:

Materials:

Three different polylactides were processed by three different ways /Fig. 1/ and exposed to an accelerated degradation test. The degradation test was evaluated by the loss of substance and of molecular mass and the loss in strength. A fiber material from an unknown source, here called PLLA-X, was declared to be a pure poly-L-Lactide. The inherent viscosity of about 3.5 dl/g was measured by capillary viscosimetry.

Parts of the fibers were solved in chloroform and then processed to nonwovens and sheets.

H. Planck M. Dauner M. Renardy (Eds.)
Degradation Phenomena on Polymeric Biomaterials

Fig. 1 Polymers and Processes

Polymer	Source	Form	Inherent viscosity [dl/g]	Samples	Titer/ Thickness	Processing
PLLA-X	unknown	Fibers	3.5	Fibers	920 dtex	Solution spun
				Nonwoven	0.30 mm	Fiber sprayed
				Sheet	0.06 mm	Knife coated
PLLA L208	Boehringer Ingelheim	Chips	2.4	Fibers	58 dtex	Melt spun
		Chips	1.9	Sheet	0.11 mm	Knife coated
		Fibers	1.2	Sheet	0.20 mm	Knife coated
PDLA R 207	Boehringer Ingelheim	Chips	1.8	Nonwoven	0.60 mm	Fiber sprayed
				Sheet	0.10 mm	Knife coated

The poly-L-Lactide L 208 from Boehringer/Ingelheim with an inherent viscosity of 2.4 dl/g was spun to fibers using an extrusion equipment. Due to the thermal and the shear stresses the material degraded to a remaining inherent viscosity of 1.2 dl/g. To produce sheets raw material L 208 of another polymer charge with an inherent viscosity of 1.9 dl/g was solved in chloroform. The close comparability of two structures needs the identic polymer. Therefore extruded fibers were solved in chloroform and formed to a sheet also.
The poly-D,L-lactide R 207 with an inherent viscosity of 1.8 dl/g was processed from a polymer solution to sheets and by air spraying to a nonwoven.

Processes:
By means of scanning electron microscopy it can be seen from the surface of the PLLA-X with the deep grooves that it must be processed by solution spinning /Fig. 2/. Above the melt spun fiber shows a very smooth surface. Please notice the different magnifications; the solution spun fiber has a diameter of about four times of the melt spun fiber!
The L 208 was spun at the Institute of Textile Technology and Chemical Engineering (ITV) using a fiber extrusion line forming the fibers through 0.5 mm nozzles. The melt temperature was about 190°C.

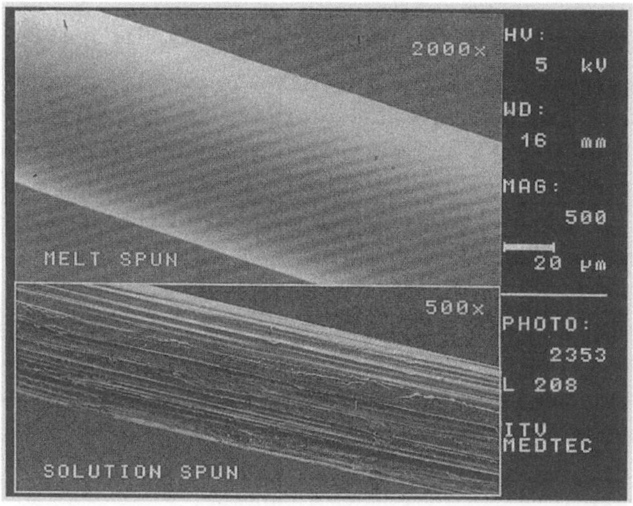

Fig. 2 Melt Spun (above) and Solution Spun (below) Fibers

Fig. 3 Air Sprayed Nonwoven

The nonwoven made of the PLLA-X and the PDLA polymer was produced using an air spraying equipment, which was developped at the ITV for processing polyurethanes to microporous structures /1/. They are used for small caliber vessels and tracheal prostheses. By means of pressurized air stream a polymer solution will be sprayed through a nozzle. The rapid drop of the pressure at the outlet of the nozzle results in an evaporization of the solvent. The polymer, if appropriate, forms short fibers which will be transported by the airstream to a rotating mandrel. They adhere together due to the remained solvent. Fig. 3 shows the scanning electron micrograph of such an microporous nonwoven.

The sheets were produced from polymer solution. The solvent is allowed to evaporate at room temperature in a dry atmosphere for 24 hours. Then the remaining traces of the solvent must be removed at vacuum at 50°C.

Degradation Test:
Preparing the samples for the degradation test they were dried at vacuum and 50°C for at least 12 hours. They were weightened and then immersed in a Sörensen phosphate buffer with a pH of 7.4. The amount of materials was depending on the structure and ranged from 0.2 to 1.4 grams. The degradation test was performed at 80°C in a recirculating air oven. The periods were from 10 hours to 23 days.
Following the respective period the samples were taken out of the buffer solution, filtered from the solution if necessary, and then washed twice in decalcinated water. After drying at vacuum, 50°C over 12 hours the samples were weightened again. The buffer solution was checked regarding the constance of the pH.

Evaluation Methods:
The degradation of the polymers was regarded with respect of the loss of substance mass, the decrease of the molecular mass and the reduction of the strength. As a measure of the molecular mass the inherent viscosity was determined in Chloroform at 25°C using an Ubbelohde Type 0C. The tensile test was performed on a universal tensile tester with a speed of the crosshead of 250% per min.
The measurement of the pH showed a constant value of about 7.4 with a small variation of +/- 0.2 which may not have effected the degradation kinetics.

RESULTS:

Molecular Mass:

For the inherent viscosity the view on the total of the investigated materials /Fig. 4/ shows three groups of curves representing the three polymers: first, the PLLA-X with its relative high molecular mass starting at an inherent viscosity of 3.5 dl/g. At the third day remains 50 %, at day 6 about 25% of the initial value. Very closely follows the PLLA L 208 as the second group with an initial viscosity of 0.9 to 1.6 dl/g. At the third day less than 30% remained, followed by a long period of seemingly constant values. Finally clearly seperated the curves of the PDLA R 207, which show a decrease of the inherent viscosity from 1.4 dl/g to near 0 at the third day.

Now looking at the PLLA-X a similar behaviour of the three different structures, the fiber, the nonwoven and the sheet can be observed /Fig. 5/. The difference at day three doesn't continuing at day six. At the third day an influence of the high crystallinity of the fibers and the large surface of the nonwoven seem to cause different degradation rates. From the sixth to the tenth day the difference is equalized.

For the L 208, again the different structures do not show a significant diverse behaviour in degradation /Fig. 6/. The sheets of the previous extruded fibers had a lower molecular mass at the beginning, but at the day six the three curves stay at the same low value.
A nearly identical curve can be regarded of the both the nonwoven and the sheet of the poly-D,L-lactide R 207 /Fig. 7/. Even though the primary molecular mass is slightly higher than that of the L 208, the decrease is more rapidly.

Strength:

Regarding the decrease of strength, both the PDLA nonwoven and the PDLA sheet could not be measured. A lateron to discuss conglomeration of the samples just after 10 hours excluded the separation of a single sample.
The curves of the strength of both poly-L-Lactides the PLLA-X /Fig. 8/ and the L 208 /Fig. 9/ with their three different structures respectively are divergent clearly.

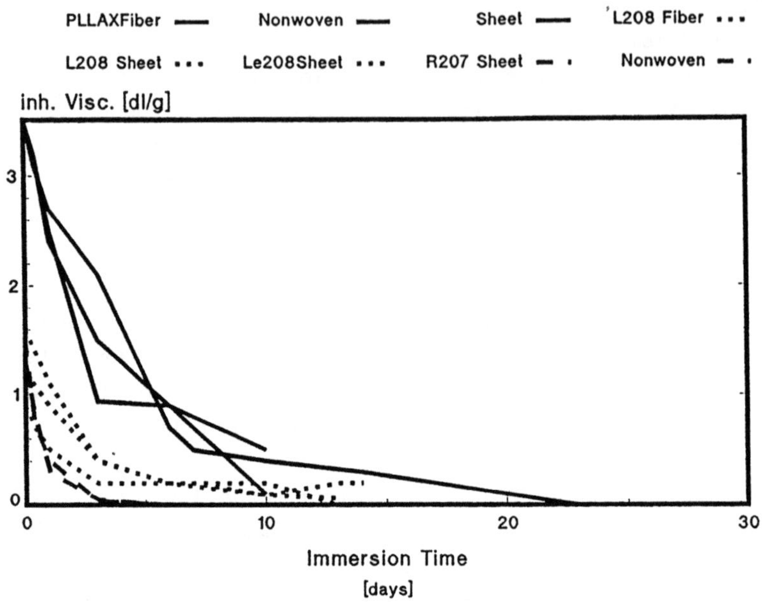

Fig. 4 Loss of Molecular Mass of Polylactides at 80°C

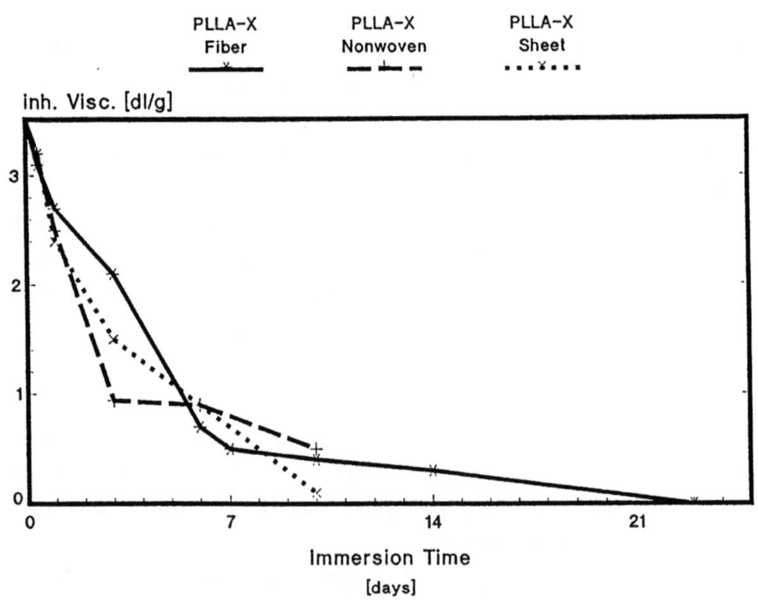

Fig. 5 Loss of Molecular Mass of PLLA-X at 80°C

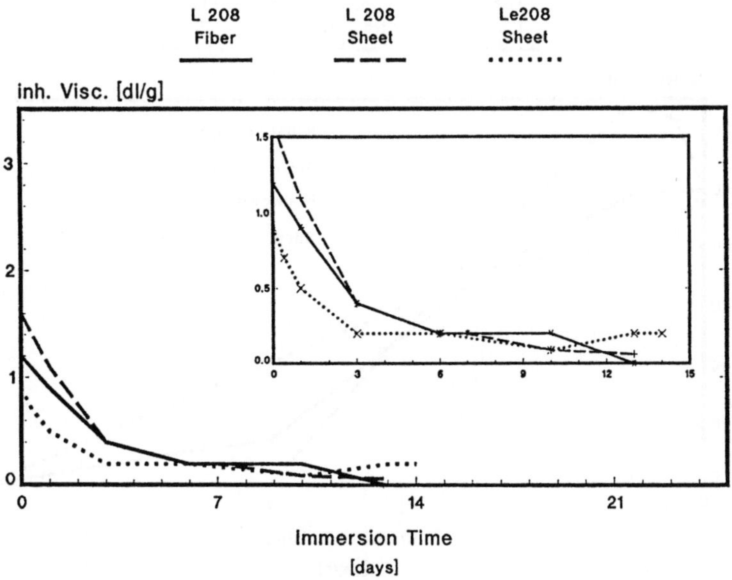

Fig. 6: Loss of Molecular Mass of L 208 at 80°C

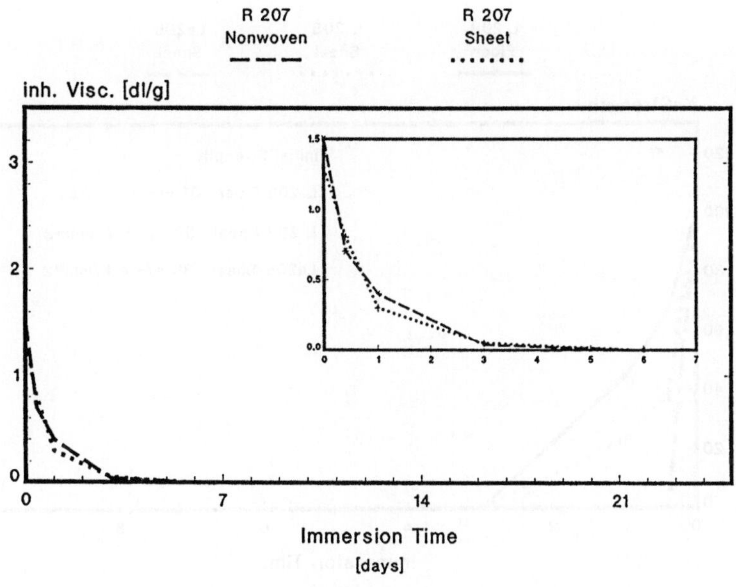

Fig. 7: Loss of Molecular Mass of R 207 at 80°C

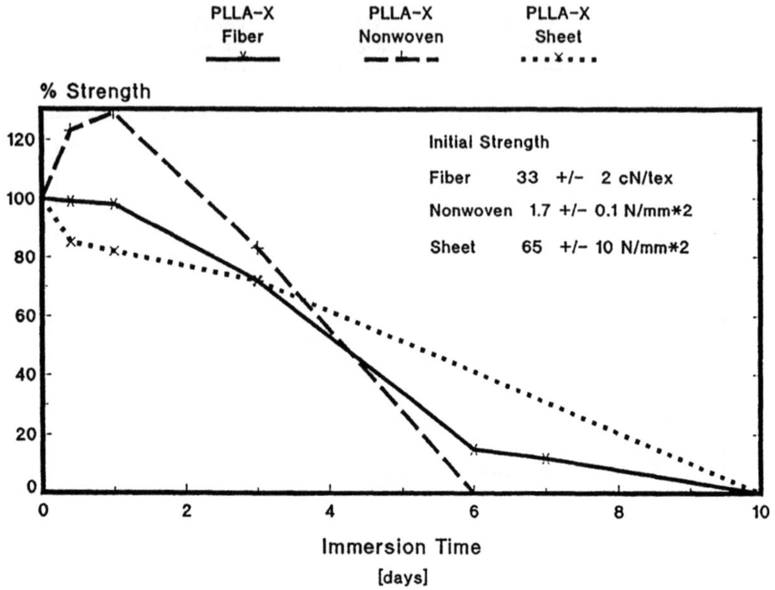

Fig. 8 Loss of Strength of PLLA-X at 80°C

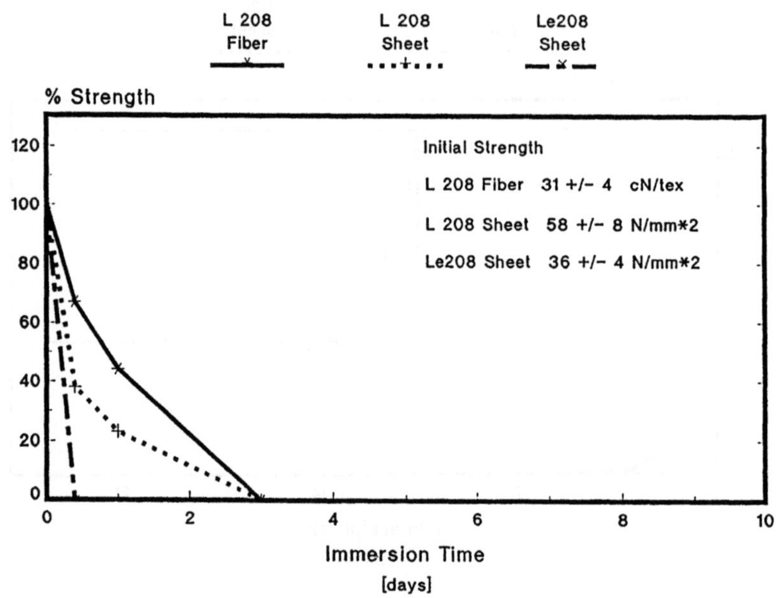

Fig. 9 Loss of Strength of L208 at 80°C

While the L 208 with its lower molecular mass has lost all strength on the third day the PLLA-X retained still 75% of the initial value.
Yet the respective courses of the decrease of strength of the different structures of the PLLA-X are not easily to understand. While the fiber material shows a classical S-shaped decrease in strength, the strength of the nonwoven increases from the beginning to the first day up to 129 %. Even though only the small number of 5 samples per period were available and the variation coefficient is about 20, the increase seems to be significant.
A reason for this surprising behaviour can only be guessed. A potential increase of the crystallinity, which enhances the strength of the fibers, is not valid for the nonwoven, because the strength is limited rather by the adherance of the short fibers than by the strength of the short fibers themselves. But an increase of that adherance may be possible, if only traces of the solvent remained from the processing. They may be diffused from the polymer at the elevated temperature and glued the fibers together. A degradation will start again at these junctions and lead to the observed fast breakdown of the strength in the following periods.
The decrease in strength of about 15% just after 10 hours which must be observed for the sheet may be due to a deformation of the sheets at the elevated temperature. The material is at room temperature below the glass transition and in the dry measuring condition brittle. Between the third day and the ten day period we had no sample available for the PLLA-X sheet. The loss in strength may run faster like with the fibers or the nonwoven.
With the L 208 a faster decrease in strength can be seen for the sheets compared to the fibers. That may be due to their lower crystallinity but despite their higher molecular mass. Because of their thickness the samples of L 208 were not deformed considerably. Obvious again is the effect of the different molecular mass as can be seen from the sheets of the original material and of the previous extruded material.

Substance Mass:
Also for the curves of the loss of substance mass a different kinetic can be seen for the three different polymers but with two deviations /Fig. 10/. The nonwoven made of the PLLA-X seems

to have a behaviour coincidental with that of the structures made of the L 208. The evidence is effected by a single value of the period of ten days. The large surface to volume relation of the nonwoven may be the reason for an accelerated degradation. But because the inherent viscosity for the PLLA-X nonwoven is identical from day six to day ten with that of the PLLA-X fibers an error in the measurement of the substance mass becomes suggestive.

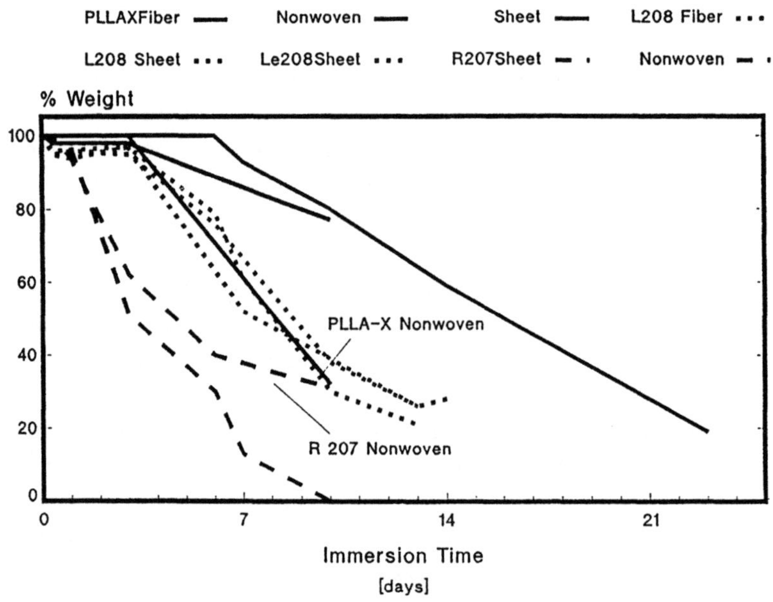

Fig. 10 Loss of Substance Mass of Polylactides at 80°C

By two consecutive values the nonwoven of the PDLA R 207 approaches in its degradation behaviour that of the structures of the L 208. Now as already mentioned with the measuring of strength, samples of both structures of the PDLA, the sheet and the nonwoven, conglomerated under the test conditions. Fibers could not be recognized any more with the nonwoven just after 10 hours of immersion. The inclusion of the phosphate salts of the buffer solution in the conglomerated structure of the nonwoven may be possible. The verification of this hypothesis was not possible. The conglomeration of both structures, the sheets and the nonwoven were surprising at first. The melting temperature of the D,L-lactide monomer is about 120° - 130°C and so significantly higher than that of the L-lactide monomer, which is about 98°C. The melting region of the poly-L-Lactide was measured to be about 175°C. So we had no adverse effect on the PDLA-polymer expected at the 80°C. Yet by measuring the softening point of the amorphous PDLA using a Kofler heating device 80° - 90°C resulted and explains the observation.

Otherwise the loss in substance mass for the different structures of the respective polymers is in accordance with the loss of molecular mass. The dependance from the molecular mass and the polymer composition is evident. The crystallinity seems to be without effect.
For the PLLA-X the %-values of the fibers and the sheets are nearly identical. We regret again the lack of the 6 day and 24 days values for the sheet.
The curves of the L 208 are well corresponding too. Yet a dependance on the small difference in the molecular mass cannot be detected. Even though the thickness of the sheet of the previous extruded polymer is twice that of the other sheet, an earlier start of the loss of substance mass may be suggestive due to the lower molecular mass.

Comparison of the Degradation Kinetics:
Now to compare the decrease of molecular mass, of strength and of substance mass for one material, two samples shall be discussed which don't show any irregularities in their curves and with a sufficient number of data. Therefore the fibers of the PLLA-X /Fig. 11/ and the sheets of the PLLA L 208 /Fig. 12/ are best suitable.
The inherent viscosity, that means the molecular mass of the PLLA-X fibers decreases from the first period rapidly to 14 % after seven days. Then a retarded decrease to 23 days can be observed. The 23 days value couldn't be measured because of the small amount of remained material. The initial strength last over the first day. The next five days the strength decreased rapidly to about 17%, followed by a retarded decrease to zero at the 10 days period. The substance mass however remained uneffected over six days. The resorption kinetic is linear til the 23 days period with 20% remained substance.

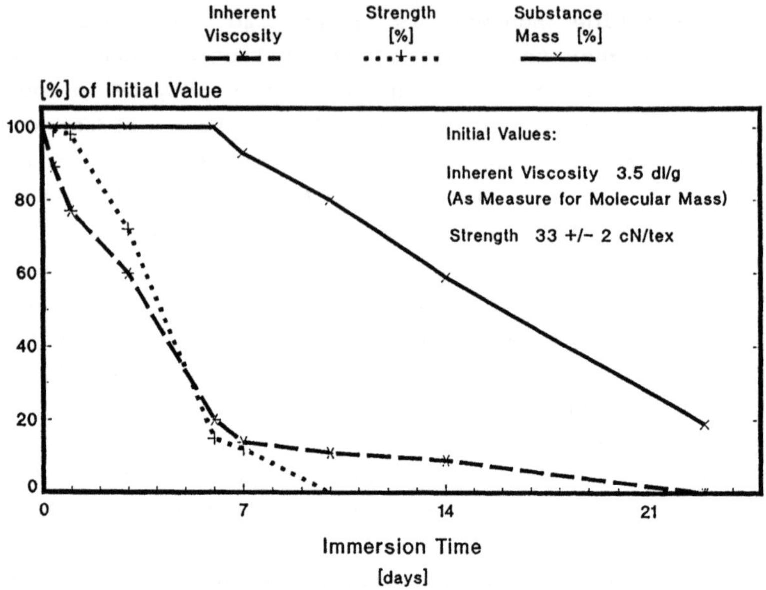

Fig. 11 Degradation of Fibers of PLLA-X at 80°C

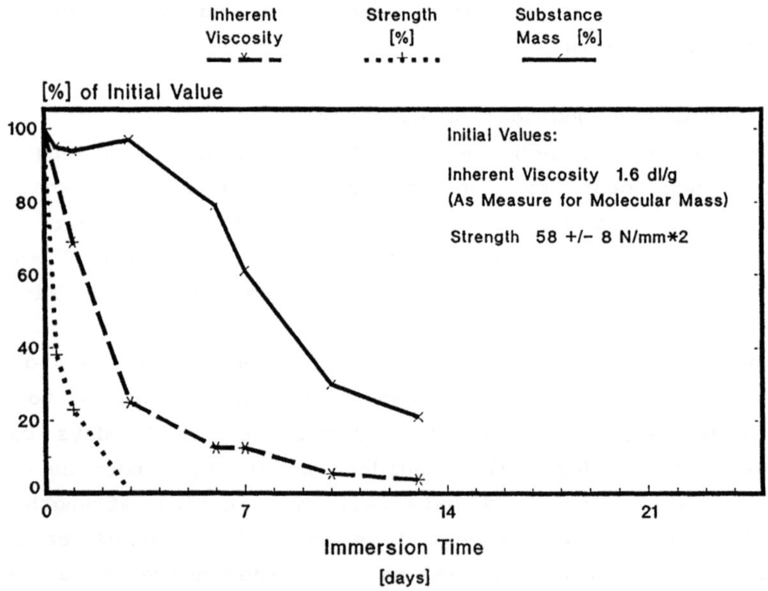

Fig. 12 Degradation of Sheets of L 208 at 80°C

If we compare the parameters it is remarkable that for 24 hours the strength is uneffected by the loss of molecular mass. Then a nearly parallel degradation kinetic can be observed til the total break down of the mechanical strength. At this time the desintegration of the substance starts.

With the sheet produced of the poly-L-Lactide L 208, the molecular mass decreases at once again, yet more rapidly, then the PLLA-X fibers does, to 25% at the 3 days period. The retarded decrease could be observed til the 13th day. Contrary to the PLLA-X fiber the L 208 sheet looses its strength from the beginning. The last value could be measured after 24 hours being 23% of the initial strength. The resorption starts at the third day to 21% of the initial mass at 13 day period.

Comparing both of the diagramms we can see at the accelerated 80°C test with polylactides that the degradation in the sense of loss of molecular mass starts at once, accompagnied with the decrease of strength. At about 20% of the initial molecular mass, or eventually an inherent viscosity of about 0.5 dl/g no mechanical strength remained. At this time the further decrease of the molecular mass is retarded, the loss of substance mass starts.
We like to point out that no mechanical strength is left at the beginning of the loss of substance mass. The desintegration of the device will start by a break of the structure prior to the real resorption, if only small forces are applied to the device at this stage. In general this results in an increase of the surface, which may help for a fast resorption. Yet for some implants in critical applications it may become problematic, when particles of relevant sizes can be put away from the implantation site and may be transported into joints or inside of blood vessels.

DISCUSSION:
It is wellknown that with an increasing molecular mass the degradation is retarded. The PLLA-X has a significant higher molecular mass than the L 208. Here the results are coincidental

with the expectation. The difference in the molecular mass of the L208 and the previous extruded Le208 can only be detected at the strength values /Fig. 13/.

Fig. 13 Expected and Measured Degradation Kinetics of Polylactides

Degradation Rate	At Increasing Parameter		PLLA-X			L 208			R 207	
			Fiber	Nonwoven	Sheet	Fiber	Sheet	e-Sheet	Nonwoven	Sheet
↓	Molecular Mass	↑	↑↑	↑↑	↑↑	O	O	↓	O	O
↓	Molecular Structure	L	L	L	L	L	L	L		
↑		D,L							D,L	D,L
↓	Crystallinity	↑	↑↑	O	↓↓	↑↑	↓↓	↓↓	O	↓↓
↑	Surface Area	↑	↑↑	↑↑	O	↑↑	O	O	↑↑	O
Measured Degradation Rate			↓	↓	↓	↑	↑	↑	↑↑	↑↑

Regarding the molecular structure the more regular poly-L-Lactide should degrade slower than the poly-D,L-lactide. And that is what we found, if we compare the L208 and the R207.
The degradation starts earlier at structures of lower order, so a high crystallinity should have a retarding effect. When we look at the different structures of the PLLA-X the fibers must have a significant higher crystallinity than the nonwoven and the lowest should have the transparent sheets. Yet no effect can be seen on that regarding molecular and substance mass. With the L 208 only at the strength values differ between the fibers and the sheets.

Finally an enlarged surface exposed to the immersion liquid should accelerate the degradation. That means nonwovens for example should degrade faster than the sheets. For the substance mass of the PLLA-X it seems to be valid. But due to the contrary kinetic of the degradation of the molecular mass this single value should not be taken in account.

While the influence of the molecular mass and the molecular structure results in the dagradation behaviour as expected, no real significant effect can be seen from the crystallinity and the surface to volume relation. This may be due to the low number of samples. That both parameters should have balanced each other that is improbable.

Besides a lot of questions which remained unanswered due to a small number of samples and too long gaps between the measuring periods, one question on which we will concentrate in the future work is the validity of the 80°C test for these materials.

The instant decrease of the molecular mass for all materials and the lack of any effect of the crystallinity is unusual. It has to be checked if beyond the glass transition of about 80°C the crystallinity for example plays a less important role than at 37°C. For the PDLA with its low softening temperature the 80°C test is unpractical. The same is to expect for all PDLLA - Copolymers. And at last, no significant effect of the surface exposed to the immersion medium could be observed. May it be possible that the diffusion rate is increased largely at the elevated temperature?

The authors like to thank Boehringer Ingelheim AG for the deliverance of the raw materials L 208 and R 207.

REFERENCE:

/1/ Patent DE 2857925

SELECTED DISCUSSION REMARKS:

Kricheldorf: You have reported about the lacking influence of crystallinity. My question is: Have you measured the crystallinity?

Dauner: No, we didn't measure the crystallinity. But the fibers - the solution spun as well as the melt spun fibers - should have a relatively high crystallinity, without knowing, which grade of crystallinity, and the sheets will have nearly no crystallinity. So we can devide this in two groups. We have no imagination of the crystallinity of the non-woven fibers.

Kricheldorf: I wouldn't agree with you, because we have looked at the rate of crystallization of poly-L-Lactide. So we see that when we quench a polylactide by precipation from solution or rapidly from the melt, the crystallization is more or less suppressed. And when we heat it slowly, crystallization is quite rapid at temperatures above glass transition. You have in your degradation experiment 80°C. Even if you have a sample which is non-crystalline in the beginning, it will crystallize within 10 minutes at 80°C. So I wouldn't expect any difference or at least no significant difference in crystallinity for your sheets and your fibers. And your assumption is not obvious, you should really measure crystallinity by Differential Scanning Calorimetry.

Dauner: Okay, we will do so, but the sheets were absolutely transparent also after immersion, and I would expect that the transparency would disappear if we had a crystallization effect.

Buchholz: In our test we used samples which we predried at 80°C. This was just to obtain a high degree of crystallinity. In some pretrials we also tested amorphous L-PLA and measured crystallinity after certain contact with water of 80°C. And in fact you obtain crystallization under these conditions.

Dauner: But this would be a clear argument against the accelerated degradation tests, for example of non-crystalline sheets, at such elevated temperatures.

Comment of the authors:

According to the proposals the samples of PLLA-X and L 208 were measured by DSC. With the PLLA-X (higher molecular weight) the degree of crystallinity ranges from approx. 68 % at the fibers and 53 % at the nonwoven fibers to 34 % at the sheets and was only slightly affected by the immersion in saline at 80°C over 24 hours. The crystallinity of the sheets was enhanced to 37%. For the L 208, the fibers had 66 % crystallinity after 10 hours (no zero value). With the sheets of L 208 the effect of recrystallization was obvious and raised up the crystallinity from 42 % to 55 %.
That confirms the observation from Fambri et al., that the recrystallization depends on temperature and molecular weight.
The recrystallization temperature shown in the DSC for the PLLA-X was at about 100°C, for the L 208 at about 106°C.

TISSUE RESPONSE TO PRE-DEGRADED POLY(L-LACTIDE)

F.R. Rozema[1], R.R.M. Bos[1], G. Boering[1], A.J. Nijenhuis[2],
A.J. Pennings[2], H.W.B. Jansen[3] and W.C. de Bruijn[4]

[1]Dept. of Oral and Maxillofacial Surgery, University Hospital Groningen, Oostersingel 59, NL-9713 EZ Groningen, The Netherlands
[2]Dept. of Polymer Chemistry and [3]Histology and Cell Biology, University of Groningen
[4]AEM Unit, Clinical Pathological Institute I, Erasmus University of Rotterdam

INTRODUCTION

Poly(L-lactide) is a member of a group biodegradable aliphatic poly(lactone)esters which have found important use as biomaterials e.g. in prosthetics, sutures, and drug delivery systems.

As-polymerized, high molecular weight poly(L-lactide) (PLLA) as synthesized according to Leenslag et al.[1] has been successfully used for fracture fixation and orbital floor reconstruction in animals and humans[2-5]. PLLA is considered to be a fully resorbable, biocompatible material[6].

In the literature on PLLA and related aliphatic polyesters, different mechanisms of biodegradation are described. Most of the studies indicate that degradation of the PLLA has to be considered as a hydrolytic process. Williams et al however, suggest also an additional enzyme-accelerated component as part of the extracellular degradation process. The final part of the absorption process is believed to happen by enzymatic surface erosion or internalization (i.e. incorporation in cells).[9-11]

In our study on the late tissue response of PLLA bone plates in humans, 3 of the 10 treated patients showed mild intermittent swelling of the implant area after a period of approximately 3.5 years. The presence of particles smaller than 10 μm, incorporated in fibroblasts which showed a 'foamy' appearance in light microscopy, and numerous intracellular poly(L-lactide) crystallites in electron microscopy was correlated with the presence of the swelling.

Paper held at the 4th International ITV Conference and Biomaterials Degradation Phenomena on Polymeric Biomaterials, September 3-5, 1991. Denkendorf, F.R.G.

H. Planck M. Dauner M. Renardy (Eds.)
Degradation Phenomena on Polymeric Biomaterials

In another of our studies on rats after 143 weeks of implantation light microscopic analysis showed macrophages with a foamy appearance[6]. The clinical manifestations as observed in the human study, could not be detected in the rat study.
High molecular weight as-polymerized PLLA takes a long time to resorb. The aim of the present investigation was to develop short lasting animal experiments in which insight in the final cellular degradation process of the as-polymerized PLLA could be obtained.

MATERIALS AND METHODS

Preparation of pre-degraded poly(L-lactide): The high molecular weight as-polymerized PLLA with a viscosity average molecular weight of 950,000, was synthesized according to the method described by Leenslag et al.[1]
To create a pre-degraded PLLA the original synthesized PLLA was machined into thin chips under cooling which were refluxed for 30 hours in a 15% solution of lactic acid in distilled H_2O. During pre-degradation-hydrolysis the PLLA disintegrated into particles with dimensions smaller than 500 μm. No sieving was performed. The obtained powder was washed several times with distilled H_2O until a pH of 6-7 was reached.
Characterization of the pre-degraded PLLA before and after the sterilization was carried out by GPC. Also, thermal analysis (DSC) was performed.
To investigate the tissue reaction to different methods of sterilization, half of the total amount of the powder was sterilized by means of regular hospital steam sterilization procedures; the other half was sterilized by gamma-irradiation of 1.8 MRad.
To examine the biological performance of PLLA particles the two types of differently sterilized powders were divided in: a portion placed into empty gelatine capsules as dry powder simulating densely packed implantation areas; and another portion, simulating scattered PLLA, which was mixed up to saturation with hydroxypropyl-methylcellulose in Ringerlactate solution (used as injection vehicle).
The gelatine capsules were sterilized by exposure to 1.8 MRad gamma-radiation.

Animal study: 14 male Wistar Albino rats were anaesthetized. Aseptic surgical technique was used for skin incisions which were made on 6 locations on the back of the rat.
In each rat separately mixed HPMC-gel and gelatine capsules filled with steam sterilized and gamma-irradiated powder, were placed in bluntly created subcutaneous pockets. As a

control, additionally one empty gelatine capsule and one portion of pure HPMC-gel were placed. Regularly clinical follow-up was planned up to 16 weeks. Two rats were sacrificed at 48 hours, 3 days and 1, 2, 4, 8, 16 weeks following the operation.
After localization of the implantation sites the tissue was generously excised and sectioned for light microscopic analysis.
Lead by light microscopic findings ultrastructural analysis was performed. From selected areas of interest sections were made.
Fixation treatments were applied to the sections in which also a cytochemical-reaction-procedure to detect the enzyme acid phosphatase was performed.
The untreated ultrathin sections were observed by transmission electron microscopy. The presence of an in-column electron energy-loss spectrometer in such a TEM instrument allows in addition to the acquisition of contrast-related images, elemental distributions images to be obtained.[12-14] This type of analytical image analysis was used to identify the acid phosphatase activity related cerium precipitates in lysosomes.

RESULTS

DSC measurements of the untreated PLLA chips revealed a heat of fusion of 56 Jg^{-1}. The pre-degraded PLLA powder showed a heat of fusion of 86 Jg^{-1}. and the number-average molecular weight was 5500 with a weight-average/number-average molecular weight ratio of 6. After sterilization only negligible differences of crystallinity and molecular weight could be determined.
Light microscopic findings showed no difference in host response between the gamma-sterilized and steam-sterilized PLLA powders. Also, no difference in tissue response was found between the HPMC-gel and the gelatin capsules insertion vehicle. There was a slight difference in distribution of the PLLA particles in the implantation area between the gelatin capsule and the HPMC-gel implants. The gelatin capsule group showed a more dense packing of the PLLA particles while the HPMC-gel group showed a loose scattering of the particles.
After 48 hours no remaining gelatine capsule or HPMC gel could be detected microscopically.
The tissue up to 16 weeks post-operatively behaved apparently homologue to experiments on polymer implants in general, comprising fibrous tissue encapsulation, lining of

macrophages and giant cells covering the smaller individual particles smaller than 10-20 μm.

The 16 week samples showed the more prominent particles surrounded by macrophages. In the mass of the large particles invasive cells were seen to occupy the spaces in between. Fibrocytic cells, macrophages and giant cells were regularly observed between extracellular connective-tissue components. When observed under crossed Nicol prisms it was noticed that at some places birefringent material, represented by the character P, was present, apparently inside macrophages and in some giant cells (Fig. 1).

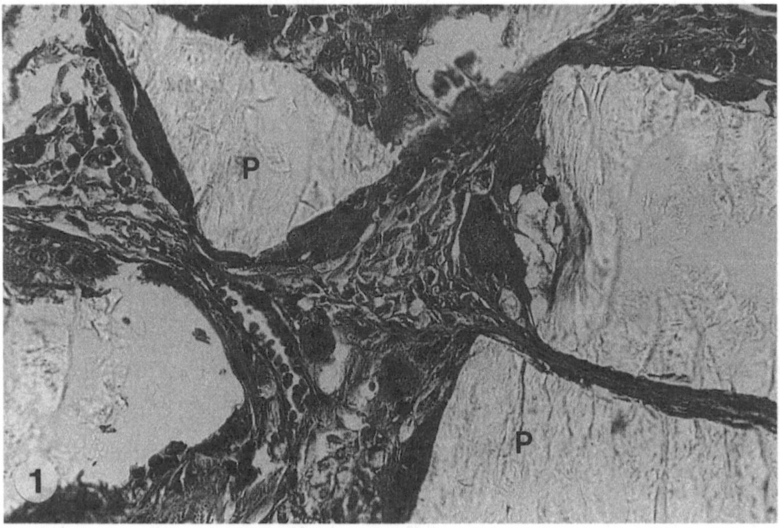

Fig. 1 Low power micrograph of PLLA particles (P) under crossed Nical prisms, showing the invasive cells and connective tissue components. Original magnification x 10, H&E stained paraffine section.

Ultrastructural sections show that in the area close to the large PLLA particles, sheets of collagen are separated by long slender fibrocytic cells, in which no PLLA material is detected. In between this connective tissue material some true macrophages with internalized PLLA fragments are under crossed Nicol prisms observed (Fig. 2).

At the sites more away from the granulated PLLA material a more mixed cell population is encountered between sheets of extra-cellular connective-tissue material. Giant cells, monocytes, neutrophils and macrophages are observed. Inside the giant cells and

macrophages PLLA material is present as irregular-shaped particles. The PLLA material is marked by black diamonds. Numerous phagolysosomes and residual bodies are present.
In addition to the normal cytoplasmic organelles, like rough endoplasmic reticulum, Golgi zones and (some giant) mitochondria, numerous phagolysosomes and residual bodies can be seen. In some macrophages very large conglomerates of PLLA material are observed consisting of several 1-2 μm fragments inside large phagolysosomes. The arrow head points to a giant mitochondrion (m); the small triangles indicate the pale mitochondrial matrix (Fig. 3).

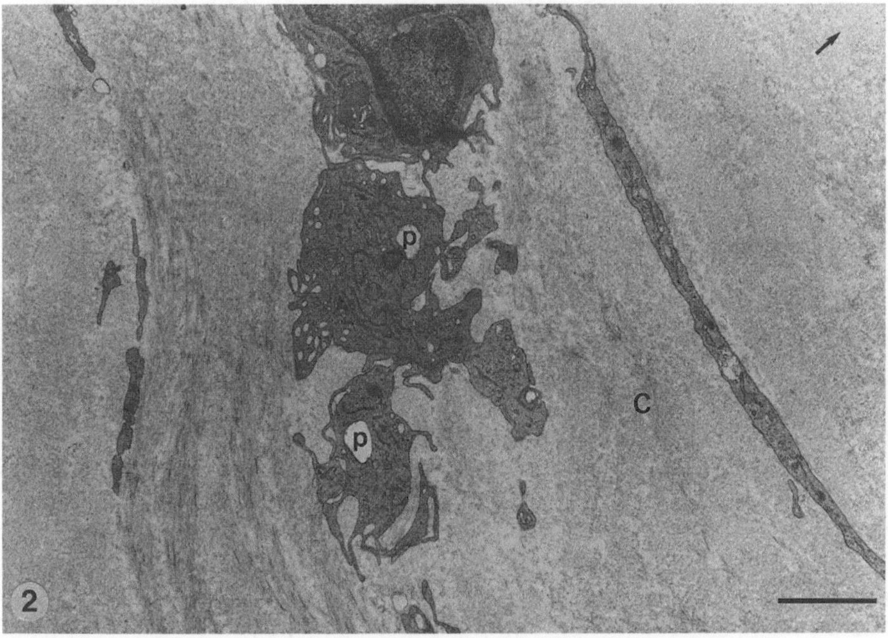

Fig. 2 Slender fibrotic cells and some macrophages amidst massive collagen sheets. The arrow points to the place of the large beam sensitive PLLA material, just outside the field of view. Bar = 2.5 μm. Uranyl acetate plus lead citrate stained section.

In some mitochondria the mitochondrial matrix is swollen and has partially lost its electron scattering capacity. In a number of macrophages numerous residual bodies are present in which small gritty fragments can be observed mixed with lipid-like osmiophilic material.
At some places multinucleated giant cells are present, in which sometimes also conglomerates of PLLA material are seen. Although most mitochondria have a normal

appearance, also in these cells some mitochondria have the empty looking matrix seen in macrophage mitochondria.

The presence of Acid phosphatase related cerium again is established in fibrocytic cells. In these lysosomes there is no sign of any internalization of PLLA fragments seen in the macrophages and giant cells at other places between the PLLA granulate material.

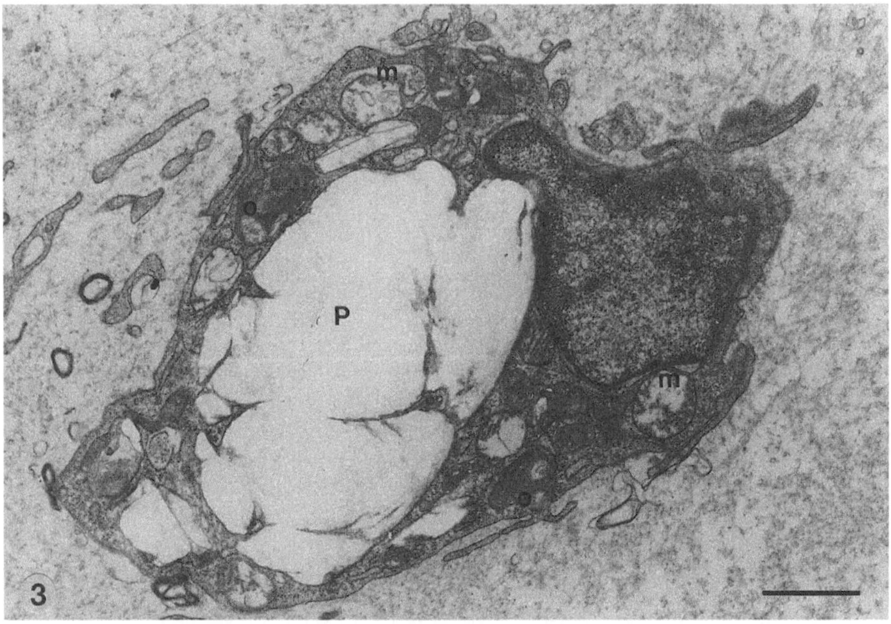

Fig. 3 In some macrophages several small PLLA fragments are agglomerated inside large phagolysosomes. Giant mitochondrion (m); the (0) indicates the pale mitochondrial matrix. Bar 1.1 μm. Uranyl acetate plus lead citrate stained section.

DISCUSSION

Until now in none of the previous **in vivo** experiments[2-5] with high molecular as-polymerized PLLA, complete resorption occurred. However, in a previous study describing the late tissue response to the as-polymerized PLLA bone plates and screws used for internal fixation of zygomatic fractures in humans,[11] the final stage of degradation and resorption seemed to be observed. This final stage of degradation caused in 4 out of the 10 patients a slight intermittent local swelling. The remnants of the PLLA bone plates and screws harvested of this stage, ca. 3 years after implantation, were characterized showing a number-average molecular weight of approximately 5000. Cellular changes and internalization of small PLLA particles were noticed which were assumed to be correlated to the final stage of degradation.

In an attempt to study long term effects of physiologically degraded PLLA in the described animal experiment, low molecular weight PLLA comparable to long term physiologically degraded PLLA[11] was made. Upon hydrolyzation the crystallinity of the material increased about 60%. The heat of fusion of the final powder was 86 Jg^{-1}. This increase in crystallinity is also observed in physiologically degraded PLLA[11] which has an ultimate heat of fusion of 96 Jg^{-1}.

GPC measurements revealed a number-average molecular weight of 5500 for the pre-degraded PLLA. These physical properties are comparable to physiologically degraded PLLA of 3.7 yrs. Therefore, pre-degraded PLLA seems to be a suitable material to study late tissue response.

The smaller PLLA particles (10-20 μm) sometimes were surrounded by giant cells. These findings were apparently homologue on a light microscopic level to the tissue response to PLLA bone plates and screws in humans after ca. 3 years.[11]

The intracellular PLLA particles as seen in the human tissue[11] and in the rat study of Bos et al.[6] were also seen light microscopically in the present study. The occurrence of internalized particles may be a sign of the final stage of degradation of PLLA. However, electron microscopical findings showed differences of appearance of the internalized PLLA particles. The internalized lamellar packed crystals present in the phagosomes of fibrocytes in the human study[11] do not resemble the internalized PLLA particles of the present study. Probably, the phase of internalized 'single crystal' polymers is not yet present.

CONCLUSION

In the present study the cells with internalized PLLA particles are well recognizable as macrophages. These macrophages contain many active lysosomes, phagolysosomes and residual bodies. At day 6 already particles from 80 down to 10 μm were phagotysed within the cytoplasm of the macrophages and giant cells. The smaller particles were within the phagosomes of the activated macrophages. Enzymatic degradation with the help of lysosome-derived acid hydrolyses is suggested.

In giant cells containing internalized PLLA particles, several mitochondria with a swollen mitochondrial matrix could be observed suggesting active digestion of lactates.

In contrast with the fibrocytes seen in the human tissue, the fibrocytes in the present study do not contain PLLA particles. However, the multinucleated giant cells also contain relatively little PLLA material.

In the present implantation study pre-degraded PLLA was internalized in a few macrophages, pointing at a late cellular degradation. The deviating aspect of the internalized particles however did suggest that the final degradation phase of the PLLA had not yet been encountered. In spite of these findings, we may yet conclude that by pre-degrading PLLA, the observation period of the final degradation phase could be successfully decreased from ca. 3.5 years to months.

Studies are in progress now in which pre-degraded particles are delivered with smaller polymer integrity in order to get more insight especially in the final cellular phase of the degradation process.

REFERENCES

1. Leenslag JW, Pennings AJ, Bos RRM, Rozema FR, Boering G (1987) Biomaterials 8:70
2. Bos RRM, Rozema FR, Boering G, Nijenhuis AJ, Pennings AJ, Jansen HWB (1989) Br J Oral Maxillofac Surg 27:467
3. Bos RRM, Rozema FR, Boering G, Nijenhuis AJ, Pennings AJ, Verwey AB (1989) Int J Oral Maxillofac Surg 18:365
4. Bos RRM, Boering G, Rozema FR, Leenslag JW, Pennings AJ, Verwey AB (1987) J Oral Maxillofac Surg 45:751
5. Rozema FR, Bos RRM, Pennings AJ, Jansen HWB (1990) J Oral Maxillofac Surg 48:1305
6. Bos RRM, Rozema FR, Boering G, Nijenhuis AJ, Pennings AJ, Verwey AB, Nieuwenhuis P, Jansen HWB (1990) Biomaterials 12:32
8. Salthouse TN, Matlaga BF (1976) Surg Gynecol Obstet 142:544
9. Williams DF, Mort E (1977) J Bioeng 1:231

10. Williams DF (1979) Spec Tech Publ 684:61

11. Rozema FR, de Bruijn WC, Bos RRM, Boering G, Nijenhuis AJ, Pennings AJ (1991) Resorbable poly(L-lactide) bone plates and screws. Tests and applicattions, Thesis, University of Groningen, Groningen, The Netherlands

12. Dort van JB, Zeelen JPh, Bruijn de WC (1987) Histochemistry 87:71

13. Sorber CWJ, de Jong AAW, den Breejen NJ, de Bruijn WC (1990) Ultramicroscopy 32:55

14. Sorber CWJ, van Dort JB, Ringeling PC, Cleton-Soeteman MI, de Bruijn WC (1990) Ultramicroscopy 32:69

MODERN ABSORBABLE SUTURE MATERIALS AND BIODEGRADABLE ANASTOMOTIC RINGS IN GASTROINTESTINAL SURGERY

A. Thiede, B. Lünstedt

Chirurgische Universitätsklinik Würzburg (Head: Prof. Dr. A. Thiede)
Josef-Schneider-Str.2, D-8700 Würzburg, Western Germany

The suture material has an important role in the every day work of the surgeon and has to face a permanent change in demands because of developments in surgery. There is no ideal suture fitting for all surgical purposes, as there is a diversity in the demands concerning the use of surgical materials. In times of extended operative procedures, there are more differentiated requirements concerning suture materials.

A new era began with the introduction of absorbable suture materials made of polyglycolic acid (PGA). For the safe use of these materials a precise knowledge of their specific handling properties is imperative (1).

To give an overview of present suture materials and an outlook on new developments and future tendencies, it is quite reasonable to ask for the arguments for medical-technological developments in surgery.

1. Extension of surgical indications,
2. improvement in surgical techniques,
3. Improvement of practicability and function of the materials.

The suture material cannot be seen independently.

There is a close interrelation of suture material including the needle-suture combination, the surgeon and the anatomic region.

H. Planck M. Dauner M. Renardy (Eds.)
Degradation Phenomena on Polymeric Biomaterials

The surgeon's personal experience and his handling abilities will help him to control the differences in handling properties. It is well known that a lack of experience causes difficulties in managing special handling properties. The surgeon should possibly have the following qualities:

- good knowledge of the material,
- experience in handling the material,
- manual dexterity and
- experience in surgery.

It depends on the surgeon's manual skills whether the loops are formed into the correct combination of loops and whether the knots are suiting to the suture material and to the anatomical region. The surgeon must be familiar with various knotting techniques when using modern suture materials. But all of this does not only depend on the surgeon's knowledge and experience, but also on the suture material with its different characteristics and the site of application.

Different anatomical regions demand different durabilities in function of the material.

Table 1:

Site of application

Area of application:	skin, subcutis, fatty, muscular tissue
	hollow organs < serosa / mucosa
	parenchymal organs
	nerves, sense organs
	vessels
	fasciae, tendons
	bones
Situation of application:	bacterial flora (state of contamination)
	pH-conditions
	enzyme activity

Conformed handling properties are equally important as well as minimal traumatic tissue damage and good knot qualities. In any case, a sufficient durability is assured by non-absorbable suture materials with the disadvantage, however, of those materials representing foreign bodies in the tissue. The use of suture material is furthermore influenced by infection, enzyme activities of the tissue and, in case of absorbables, by the time of absorption.

The suture itself is influenced by its material, manufacturing, composition of surface, coating and needle-thread combination. Suture materials can be devided into the following classifications:

- basic substance,
- form of processing,
- surface coating,
- needle-thread combination.

We differentiate resorbable materials, for example catgut, which must be reabsorbed by an active enzyme or cellular mechanism, and absorbable materials, which are water soluble. There are ultrashort term, short term and medium term absorbable materials such as Vicryl rapid, Dexon, PDS, and Maxon.

Table 2:

Absorbable suture materials on sale

Function	Trade Name	Coating	Mono(M)-/ Polyfilament(P)	Function Time days	Absorption Time days
USAS	Vicryl Rapid	+	P	14	~50
SAS	Dexon S		P		
	Dexon Plus	+	P	~35-30	~90-120
	Bondek	+	P		
	Vicryl	+	P		
	Medifit		P		
MAS	Maxon	_	M	~ 40-60	~120-160
	PDS	-	M	~60-80	~180-240

U=ultra,S=short-term,M=medium-term,A=absorbable,S=suture material

The physical properties of the materials substantially influence handling. Among these properties are the diameter, thread structure, type of twisting, flexibility, elasticity, possible reversible elongation, tensile strenght and knot security (2).

The biochemical characteristics of the material play an important role especially concerning the synthetic absorbable materials with regard to tissue reaction, duration of function, absorption rate and handling. Due to the absorption process with its special antibacterial environment, it is possible to heal suture abscesses without needing to remove the material. This is even more advantageous when polyfile absorbables are used (3).

Table 3:

Biochemical properties

basic substance of the thread and length of polymer chains	function time absorption time
suture coating	handling
absorption process (conditions of the microenvironment)	allows healing of abscesses without surgical removal of sutures

In the group of absorbable materials there are the monofilaments with only 1 filament, polyfilaments with many filaments and the coated thread in which the individual filaments are coated, for example:

Figure 1:

Form of processing

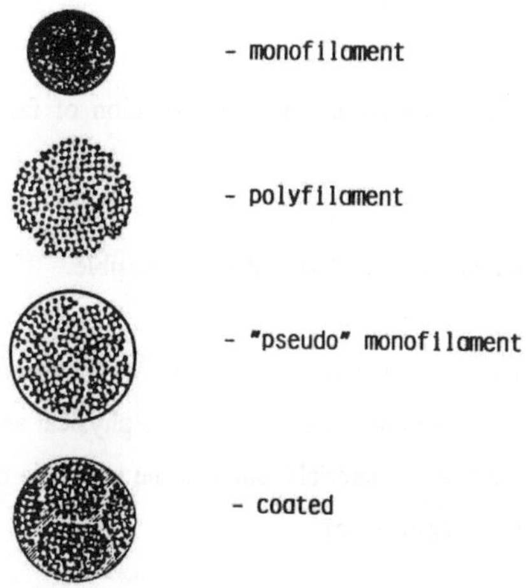

- monofilament: Maxon and PDS,
- polyfilament: Dexon 1st generation,
- coated thread: Dexon bicolor and Vicryl.

The contamination of surgical wounds with pathological germs during or following the operation is not a rarity. Only in few cases, an infection would result. A healthy wound has a surprisingly high resistance against infections. Such resistance can be negatively influenced, however, by local factors. In addition to the devitalized tissue, suture material as a foreign body plays an important role. It has been proven that by using implanted sutures, the concentration of bacteria necessary to infect a wound has been greatly reduced.

The biological mechanism of infections and the healing of wounds are basically similar and are comparable in the development of their cellular and vascular reactions.

The implanted suture also produces an almost analogical reaction. The appearance of monocytes, granulocytes and lymphocytes are typical of reactions produced by the tissues. These cells would appear in infections as well.

In the beginning stages of this reaction, granulocytes are found within the suture area. The macrophages and its derivatives, subsequently, determine the cellular formations in and around the thread.

The giant cell made up of monocytes is a characteristic cell for the fine tissue formation of foreign body reaction. This is also characteristic for suture material.

A typical reaction during the absorption of synthetic materials is the formation of fatty necrosis around the material.

The healing of an abcess after controlled contamination of a PGA suture is possible.

The properties of handling influence the behavior of the material, first by the way it passes through the tissue and second by the knot function. Both are influenced by the physical and biochemical characteristics of modern absorbable suture materials, such as the structure of the thread, surface texture, coating and flexural strength (4, 5).

For the surgeon, the function time (FT) of a suture material is very important. The definition of function time of a suture is the period in which the thread processes measurable tensile strenght. The FT 50 is defined as the moment at which an absorbable thread only shows 50 % of its original tensile strenght. Absorption time is the period in which a thread or its residue can be identified, histologically, in the tissue.

The ratio between function time and time of absorption, generally, is 1 : 3.

The graphs show the function times of Vicryl rapid, of Dexon and Vicryl, of Maxon and PDS. By comparing Maxon and PDS, we can see that Maxon has a greater original tensile strenght, whereas PDS shows a slightly longer time of function (6).

Figure 2:

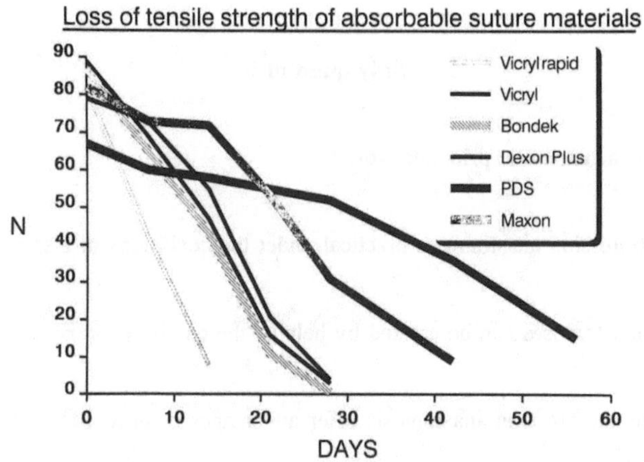

The histological absorption of the braided Dexon suture shows a continuous decrease in the suture diameter over the course of 90 days. The absorbable monofilament Maxon also shows a continuous decrease in the diameter of the suture over the course of 200 days.

In conclusion it can be said that monofilaments have a less pronounced influence in tissue reaction and infection probability.

The most recent development in biodegradable surgical instruments is the so-called Valtrac anastomotic ring. This anastomotic ring consists of fully absorbable polyglycolic acid and barium sulfate as a contrast medium. It allows the creation of inverted sero-serous intestinal anastomoses. The precursor of the modern biodegradable anastomotic rings (BAR) is the Murphy button presented by Murphy in 1892. The Valtrac ring is a modern concept of standardized anastomoses in the GI tract and the ring does not remain in situ as a foreign body, but predominantly fragments. As initial experimental and clinical studies showed promising results, the use of this anastomotic ring had to be examined under every days conditions in a surgical hospital (7, 8).

Study questions:

1. Does this anastomotic principle work?

2. Are compression anastomoses practical under the conditions of a specialists clinic?

3. Which anastomoses can be created by help of the compression ring?

4. Does the compression anastomosis offer advantages or draw-backs as against other techniques?

5. How high is the postoperative complication rate and how can these problems be controlled?

Patients and Methods

The Valtrac Ring: The Valtrac anastomosis ring is made of 88% polyglycolic acid and 12% barium sulfate. The barium sulfate component allows simple identification of the ring in the overview radiograph of the abdomen. The polyglycolic acid permits the ring to fragment into descrete parts or to be completely dissolved. The ring consists of 2 mushroom-shaped half rings with a bridge mounted on an applicator. In the open state, the distance between the half rings is 6 mm. The rings are currently available with four different outer diameters: 25, 28, 31, 34 mm. Each ring size is available in two variants with respect to the compression zone, namely 1.5, 2.0 or 2.5 mm. These figures refer to the ring distance obtaining in the adapted state and must be selected according to the thickness of the bowel wall being compressed.

Fixation forceps matched to the ring sizes are a great help in introducing the rings into the second bowel stump. They improve the practical usefulness of the Valtrac rings considerably.

The Operation: Our approach makes use of the standard preoperative preparation with orthograde bowel enemas, intraoperative ultrashort term antibiotic prophylaxes, peri- and postoperative administration of low-dose heparin, parenteral feeding and visualization of the anastomosis with water soluble contrast medium between the 7th and 8th day after the operation.

Operational Technique: The attached connective tissue in the planned anastomosis region is removed along a length of about 5 mm proximally and distally. Very good circulation in the stumps must be ensured. The anastomosis region is washed with a desinfectant solution. Any projecting mucosal edge is resected. A continuous overhand purse-string suture using Maxon of gauge 2-0, rarely 3-0, is then performed. Alternatively, a purse-string suture clip can also be applied to the small bowel. The proximal part of the anastomosis prepared in this way is then pre-extended with dressing forceps and spread over Ellis clamps or holding sutures. In this way, the Valtrac ring can be introduced with the applicator and the circular suture sewn over the bridge.

The bridge of the Valtrac ring is then fixed manually or by means of a specially designed fixation forceps. The plastic applicator can then be removed. The forceps facilitate the introduction of the ring into the distal part of the anastomosis. As soon as the purse-string suture is ready and the forceps have been removed, the anastomosis can be completed by simple compression. The holding mechanism comes to a stop inside the ring. The maximum stress of the anastomosis is now tested by the surgeon applying a slight traction to the bowel.

During the inspection, one should carefully ensure the seroserous integrity is maintained. Any projecting mucosa lips must be resected. Small serosal lacerations can be oversewn with a 4-0 or 5-0 Maxon suture.

Patients characterization: A total of **240** (180 of which have been evaluated) Valtrac rings were used during the period between March 1989 and **July 1991** (evaluation until February 91). The most frequent indication was a colon or high rectum carcinoma (66 anastomoses) followed by diverticulitis (36 anastomoses) and a sigmoid resection in connection with a rectum prolapse operation (32 anastomoses). Further, less common indications were gastric carcinoma, pancreatic pseudocysts, chronic bowel inflammations, esophagus carcinoma, large polyps of the upper rectum, gastric lymphoma and pancreas head carcinoma.

Table 4:
Valtrac Anastomoses 3/89 - 2/91

Diagnoses (patients n=164)

Diverticulitis	36
Cancer of colon/rectum	66
Rectal prolaps a. sigmoidectomy	32
Sigma elongation	6
Gastric carcinoma o. lymphoma	4
Gastric ulcer	1
Crohn's desease	4
Radiation stenosis - small bowel	2
colon	2
Cronic small bowel desease	3
Pancreas-Pseudocysts	2
Pancreas-carcinoma	3
Oesophageal carcinoma	3
Total	n = 164

A total of 107 anastomoses were created in women, 73 in men. The mean age was x = 65 years (range from 23 to 85 years).

In total, 58 end-end anastomoses were created, 22 of which were end-end colon anastomoses and 36 end-end small bowel colon anastomoses. The 20 end-side anastomoses were created largely in the course of Y-Roux shunts after gastric resections, whereas 2 side-side anastomoses were created during a shunt from the stomach to pancreas pseudocysts during a drainage operation.

Results

1. Intraoperative findings: 4 patients had to be excluded from the study because even the smallest anastomosis ring could not be used due the small size of bowel lumen. In the case of 3 patients, the anastomosis could not be created using the planned ring size and a smaller ring had to be selected. In 15 cases, small serosal alterations occurred when closing the purse-string suture. In 14 occasions a small mucosa prolapse was observed when closing the compression ring. This was resected in each case. In the case of both the serosal alteration and after treated mucosa prolapse, the defects were oversewn with 4-0 or 5-0 Maxon. All those defects were small and without leakages. Intraoperative bleeding from the anastomosis region did not occur.

Table 5:

Valtrac Anastomoses, 3/89 - 2/91

Intraop. finding a. measures

Side of Anastomosis	n	Dehisc.	Over-sewing	Fibrin sialant	Mucosal prolaps	Hemorrhages	Locking pr.def.	Mechanism Correction
Coloninterposition	1	-	-	1	1	-	-	-
Oesophagus	2	-	1	-	-	-	-	-
Stomach	6	-	1	-	-	-	-	-
Pancreaspseudocyst	3	-	-	-	-	-	-	-
Choledochus	1	-	-	-	-	-	-	-
Small intestine	20	2	2	1	-	-	-	-
Small-large-intestine	25	2	1	1	2	-	-	-
Colon	64	5	9	3	6	-	2	-
Rectum	58	6	8	5	6	-	1	-
Total	180	15	21	11	14	-	3	-

2. Postoperative findings: In a total of 180 patients, there was not any case of mortality due to the operational technique. 1 patient died 4 months after the operation from the consequences of a cardiac infarction, 1 female patient died from haemorrhagic shock 4 weeks postoperatively after arrosion of the splenic artery for pancreas necrosis. This complication did not occur when the Valtrac ring for gastrocystostomy was applied. Clinically relevant bowel blockages have not been observed. One patient already discharged from hospital was readmitted on the 18th postoperative day after sigmoid resection due to adenocarcinoma with subileus condition. Radiologcal examination showed a short spastic bowel segment with normal dilatability of the transverso-rectostomy oral of the anastomoseis. Under conservative treatment, the symptoms declined within two days.

Anastomostic leakages were observed in the colon region of 8 patients by means of routine x-ray examinations with water-soluble contrast medium between day 7 and 10 postoperatively. Only 3 out of these patients, however, showed a corresponding clinical correlate, and 2 of them also exhibited temporary colocutaneous fistulization in the drainage canal. The other 3 patients did not show clinical symptoms. In none of the patients reoperation was necessary due to the anastomotic leakage, and healing could be assured in each case by conservative treatment.

Table 6:

Valtrac Anastomoses 3/89 - 2/91

Postop. complications

Side of Anastomosis	n	Hemorrhage clin.	Hemorrhage Reop	Leakage radiol.	Leakage clin.	Leakage reop.	Ex(techn)	Fistula	Stenosis
Coloninterposition	1	-	-	-	-	-	-	-	-
Oesophagus	2	-	-	0	1	1	-	-	-
Stomach	6	-	-	-	-	-	-	-	-
Panreaspseudocyst	3	-	-	-	-	-	-	-	-
Choledochus	1	-	-	-	-	-	-	-	-
Small intestine	20	1	-	-	-	-	-	-	-
Small-large-intestine	25	-	-	-	-	-	-	-	-
Colon	64	-	-	4	1	-	-	1	-
Rectum	58	-	-	4	1	-	-	1	-
Total	180	1 0,5%	-	8 4%	3 1,5%	1 0,5%	-	2 1%	-

Postoperative bleeding from the anastomosis region was observed in one of 180 anastomoses. The dissolution and elimination of the Valtrac ring had been completed in all patients between 12 and 30 days postoperatively.

The patients did not notice the elimination of the ring fragments in 88 %. In the case of 1 patient, the complete anastomosis ring was eliminated on the 12th day without any disturbance of anastomotic healing. The 30 patients observed during the 6 - 12 months follow-up period were examined radiologically or endoscopically. Neither stenoses nor a leakage were diagnosed in any case. In 78 % of the x-ray colon follow-up examinations, the anastomosis region could not be distinguished any more from the rest of the colon.

Under endoscopic examination after elimination of the Valtrac ring, the compression anastomosis could only be seen as fine line.

Answers to the questions:

1. The anastomotic principle of Valtrac rings works excellently when taking into account some technical tricks.

2. Even under training conditions, the operatinal technique is easy to learn. So it can be successfully applied by many surgeons of the most varied levels of training, with a low rate of complications.

3. Currently, the following localizations can be recommended for Valtrac anastomoses: small bowel (E-E + E-S), ileocoecal region (E-E), large bowel (E-E), upper third of the rectum (E-E).

4. Favourable factors include the standardability of the method, and the absence of haemorrhages and stenoses. Currently, an applicator for anastomoses in the lower third of the rectum is unavailable. This anastomotic principle may prove to be

inapplicable for great variations of bowel wall thickness and for relatively narrow lumina.

5. The radiological dehiscence rate was 4 %; the clinically relevant rate was 1.5 %. Reoperation was required in one esophageal anastomosis; stenoses have not been recorded.

Discussion

The discussion about the optimal technique of enteral anastomoses still remains inconclusive despite numerous experimental and clinical studies. The behaviour of the various adaptation materials as well as the effect of the various types of anastomoses must be taken into account.

Animal studies showed that the maximum stress of bowel anastomoses declines over the first 4 to 5 days, consolidation of the bowel suture occurring rapidly only with an increasing collagen content of the bowel wall.

In this phase, which is critical for an anastomotic dehiscence, therefore, the important thing is to ensure a secure adaptation of the bowel ends. This can be achieved, on the one hand, by means of a suitable anastomosis technique and, secondly, by good preoperative preparation of the bowel. This may be achieved by an orthograde colonic irrigation, which is of decisive importance especially for colon anastomoses (9, 10).

The pathogen-reducing action of the colonic irrigation can be enhanced by the addition of non-absorbable antibiotics to the irrigation fluid as well as by systemic short-term administration of antibiotics (11, 12). By this means, microabscesses in the suture region, which could promote a deshiscence, can be prevented. Also of critical importance is the avoidance of local circulation disturbances due to too extensive skeletization of the bowel ends or the creation of anastomoses under tension.

Subtile haemostasis in the anastomosis region is an urgent necessity, as haematoma formation in the vicinity of anastomoses promote infections. Such infections lead to an increased activation of the collagenase, resulting in an increased collagen decomposition and reduced stability of the anastomosis (13).

The foreign material introduced for the anastomosis at the bowel ends leads to inflammatory environmental reactions. The scale of this inflammation varies depending on the individual materials. The reaction of the surrounding tissue is the smallest to polyglycolic acid and steel, in comparison with the rest of suture materials (14).

Indications for Valtrac Ring Anastomoses

A Valtrac ring anastomosis can be performed in the entire region of the intestine for end-end, end-side, side-end and side-side anastomoses. Its application in the region of esophagoenteral or esophagogastric anastomoses, its use for drainage of pancreas pseudocysts as well as small bowel anastomoses cannot yet be regarded as a completely secured indication. In the relatively small number of our cases, hitherto, therehave not been recorded any complications. The theoretically possible bowel obstruction via the ileocoecal valve, in practice, did not represent a relevant problem for the elimination of the ring fragments.

Neither ileus or subileus conditions nor stenoses or postoperative bleeding occurred in these anastomotic localisations.

The use of Valtrac ring anastomoses for colon and small bowel colon anastomoses is secured both by our own results (15) and by studies reported in the British and American literature. Hardy showed in his comparative study of dogs and pigs that the creation of anastomoses with the Valtrac ring is simpler and faster than by a manual suture technique or clip suture. The healing trend was also examined by microscopy and was about equally effective for all methods, although tissue necroses were least marked with the Valtrac ring, which is of great importance for the reliability of the anastomosis.

In the first clinical study, Hardy observed a complete lack of complications due to the Valtrac ring. Two major comparative group studies showed about the same complication rates in a comparison of the Valtrac ring, the clip suture device and manual suturing (16, 17). The manipulation of the Valtrac ring seemed, however, to be the simplest. In addition, the greater reliability of the compression ring for inflammatory bowel conditions or for a pre-irradiated bowel was also recorded (18, 19).

A greatly thickened bowel wall, in the case of edema or an excessively narrow bowel diameter can represent limitations to the use of Valtrac rings. This will rarely occur, however.

Conclusion

Enteral anastomoses with the biofragmentable Valtrac ring can be reccomended as offering simple-to-learn manipulation and the creation of effectively standardized enteroanastomoses at various bowel sections. Stenoses or postoperative bleeding, practically, do not occur at all. The suture dehiscence rate is comparable to that for clip suture anastomoses or manual suture anastomoses at corresponding localizations.

The results hitherto observed with Valtrac, in this prospective study, allow to draw the following conclusion: In the colon and the upper and middle region of the rectum as well as in the small bowel region, compression anastomoses will work excellently offering an attractive alternative to manual or clip sutures. This is the case, particularly, in view of the target criteria of standardazation, low complication rate and absence of stenoses. The experience gained in the esophagus, stomach and at the bile ducts, currently, is still insufficient to allow an assessment.

Ultimately, the comparative value of this method will only be assessed in controlled studies.

References

1 Thiede A, Hamelmann H (eds) (1982) Moderne Nahtmaterialien und Nahttechniken in der Chirurgie. Springer, Berlin Heidelberg New York

2 Lünstedt B, Knoop M, Thiede A (1986) Übersicht physikalsicher und handhabungstechnischer Eigenschaften moderner, monofiler Nahtmaterialien. Chirurg 57 (8) 510-513

3 Thiede A, Jostarndt L, Lünstedt B, Sonntag HG (1980) Kontrollierte experimentelle histologische und mikrobiologische Untersuchungen zur Hemmwirkung von Polyglykolsäurefäden bei Infektionen. Chirurg 51: 35-38

4 Rodeheaver GT, Powell TA, Thacker JG, Edlich RF (1987) Mechanical Performance of Monofilament Synthetic Absorbable Sutures. Am J Surg 154 (5) 544-547

5 Thiede A, Stüwe W, Lündstedt B (1989) Vergleich von physikalischen Parametern und Handhabungseigenschaften kurzfristig und mittelfristig absorbierbarer Nahtmaterialien. Chirurg 56: 803-808

6 Knoop M, Lünstedt B, Thiede A (1987) Maxon und PDS - Bewertung physikalischer und biologischer Eigenschaften monofiler absorbierbarer Nahtmaterialien. Langenbecks Arch Chir 371 (1) 13-28

7 Hardy TG jr, Pace WG, Maney JW (1985) A biofragmentable ring for sutureless bowel anastomosis. An experimental study. Dis Colon Rectum 28: 484-490

8 Hardy TG jr, Aquilar PS, Stewart WRC (1987) Initial experience with a biofragmentable ring for sutureless bowel anastomosis. Dis Colon Rectum 30: 55-61

9 Denecke H, Wirsching R (1984) Colorectale Anastomosen. Chirurg 55: 638-644

10 Marti MC, Pouret JP (1977) Eine neue Methode zur raschen präoperativen Vorbereitung des Dickdarms. Langenbecks Arch Chir 343: 275-279

11 Jostarndt L, Thiede A, Sonntag H-G, Hamelmann H (1982) Kontrollierte, prospektive, randomisierte Studie zum Wert der systemischen Antibiotikumprophylaxe in der elektiven Dickdarmchirurgie. In: Lange J, Theisinger W (Hrsg) Aktuelle Therapie des Rektumcarcinoms, Thieme, Stuttgart, pp 28-34

12 Heil K (1980) Antibiotika in der kolorektalen Chirurgie. Coloproctol 4: 227-230

13 Lünstedt B (1988) Experimentelle Untersuchungen zur Wundheilung der Kolonanastomose unter besonderer Berücksichtigung der Nahtmaterialien und des gewebeständigen Kollagenstoffwechsels. Habilitationsschrift, Medizinische Fakultät der Christian-Albrechts-Universität zu Kiel.

14 Nockemann PF (1980) Die chirurgische Naht. Thieme, Stuttgart, pp 161-162

15 Klima J (1991) Prospektive Studie über die Anwendung eines biofragmentierbaren Kompressionsringes. Konzept, technischer Einsatz, Ergebnisse, Wertigkeit. Inauguraldissertation Universität Hamburg.

16 Cahill CJ, Betzler M, Gruwez JA, Jeekel J, Patel J-C, Zederfeldt B (1989) Sutureless large bowel anastomosis: European experience with the biofragmentable anastomosis ring. Br J Surg 76: 344-347

17 Corman ML, Prager ED, Hardy TG jr, Bubrick MP, and the Valtrac (BAR) study group (1989) Comparison of the Valtrac biofragmentable anastomosis ring with conventional suture and stapled anstomosis in colon surgery: results of a prospective, randomized clinical trial. Dis Colon Rectum 32: 183-187

18 Maney JW, Katz AR, Li LK, Pace WG, Hardy TG jr (1988) Biofragmentable bowel anastomosis ring: comparative efficacy studies in dogs. Surgery 103: 56-62

19 Smith AD, Bubrick MP, Mestitz ST (1988) Evaluation of the biofragmentable anastomosis ring following preoperative irradiation to the rectosigmoid in dogs. Dis Colon Rectum 31: 5-9

SELECTED DISCUSSION REMARKS

Urhøj: I would just like to know if you could tell me what material the BAR (biodegradable anastomotic ring) was made from.

Thiede: Yes, 85 % PGA and 12.5 % contrast medium for x-rays.

Guidoin: Could you tell us, who manufactures these rings, these BARs?

Thiede: It is manufactured by Davis & Geck.

Guidoin: Is it commercially available now?

Thiede: It is available in Germany, in France, in Sweden, in Finland, and they have just started in the United States.

Nakamura: In Japan, many surgeons use Auto-sutures for end-to-end anastomosis. What is the advantage of your rings?

Thiede: Auto-suture devices are filled with staples, with metal staples, and that are foreign bodies for a very long time, for the whole life, when it is used in the intestinal localisation. There are 2 practical or theoretical disadvantages: One is the foreign body reaction. You will find in 2 - 5 % stenoses after you have used metal staples. And the second disadvantage is: The follow-up is more complicated, because the NMR-investigation is not so clear. We use Auto-suture circular anastomotic devices also in many cases. It is very practical in the very deep rectal region and also in the oesophagus, but not in the higher regions of the colon and in the intestinal tract, also in the gastric region. And in my opinion it is the best to have no residues of the suture material after you have done your work and you have waited about 3 weeks when the anastomoses are healed.

A NOVEL BIOABSORBABLE MONOFILAMENT SURGICAL SUTURE MADE FROM (ε -CAPROLACTONE, L-LACTIDE) COPOLYMER

Tatsuo NAKAMURA, Yasuhiko SHIMIZU, Teruo MATSUI, Norihito OKUMURA, Suong Hyu HYON, Kouji NISHIYA*,

Research Center for Biomedical Engineering, Kyoto University, 606 Kyoto, Sakyo-ku, JAPAN, Gunze Co.*

Abstract:

A Novel monofilament bioabsorbable sutures was made from a copolymer of ε-caprolactone and L-lactide (Poly(CL-LA)). The Composition weight ratio of L-lactide: ε-caprolactone was 85 : 15. Initial tensile strength of the 5-0 suture was 838 ±72 gf (knot-pull 600 ±60 gf). An in vivo degradation test using rabbits showed: a poly(CL-LA) suture retains its strength for longer periods than other synthetic monofilament absorbable sutures such as MAXON® and PDS.® %Retention of tensile strength at 4 week was Poly(CL-LA) 83 ± 13%, MAXON ®36 ±9%, PDS® 50 ± 15%. At 8 weeks the Poly(CL-LA) was 38.2%, though the MAXON ®and PDS®were 0%. The tissue reaction of this suture was not remarkable at any interval. At one year, the Poly(CL-LA) fiber was replaced with collagen deposition.

Introduction:

In vascular surgery continuous suturing with monofilament sutures has been routinely used. These monofilament sutures have been made from non-absorbable synthetic polymers such as polypropylene and polyethylene. In vascular surgery absorbable sutures are not so frequently used in comparison with tracheo-brochial surgery. The reason may be because cardiovascular surgeons have anxiety as to the rupture of anastomosis, which may produce fetal complications. As regards the mechanical function, sutures have only to fix the anastomoses site until the wound healing finishes. In other words, the function of the suture is accomplished when the wound healing is finished and the wound is able to withstand stress without suture support. After healing of the anastomoses, since any suture material is a foreign body for living tissues, it is desirable that they will not stay at the anastomoses site[1]. Especially in the cases of pediatric surgery, the continuous suturing with non-absorbable sutures may cause a constriction of the anastomosis portion as a child grow up.

H. Planck M. Dauner M. Renardy (Eds.)
Degradation Phenomena on Polymeric Biomaterials

The object of this study is to develop a reliable monofilament absorbable suture which can be applied for vascular surgery as a prelude to clinical trials.

Materials and Methods:

A copolymer of l-lactide and ε-caprolactone (hereinafter abbreviated as poly(CL-LA) was synthesized from l-lactide and ε-caprolactone by ring-opening polymerization. The Molecular structure copolymer is shown in Figure 1. This copolymer is a polyester which degraded by hydrolysis. Three copolymers with l-lactide to ε-caprolactone weight composition ratio 85.0:15.0, 76.0:24.0 and 73.5:26.5 were prepared. No plasticizer was added during the polymerization process. 3-0 monofilament sutures were made from these three copolymers. For these sutures, an in vitro degradation test was performed in 37 °C saline solution. The result of %retention of tensile strength is shown in Figure 2. Since the object of this study is the development of a relatively slowly absorbable suture, we chose the copolymer with composition ratio 85:15, and the following in vivo tests were performed.

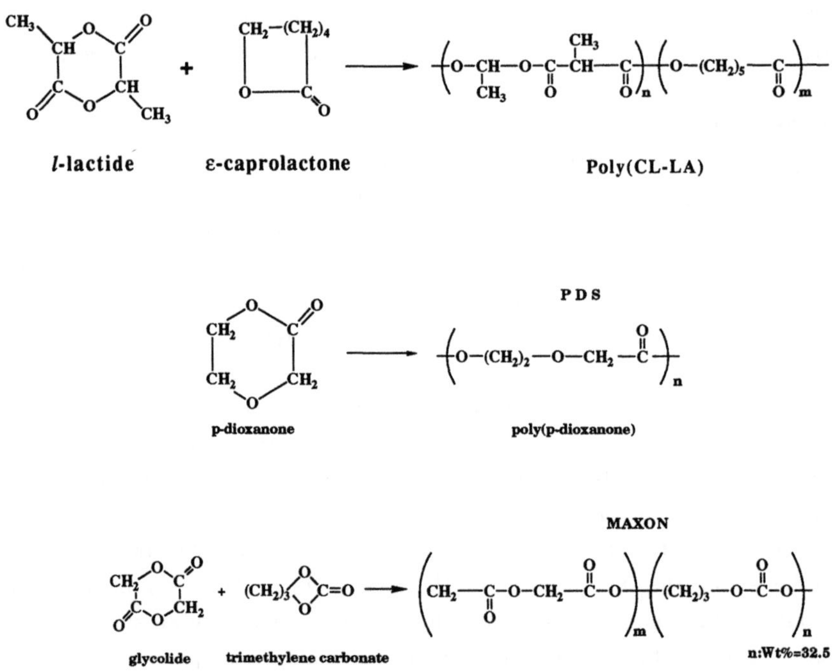

Fig.1 Molecular Structures of the Polymers for Monofilament Sutures

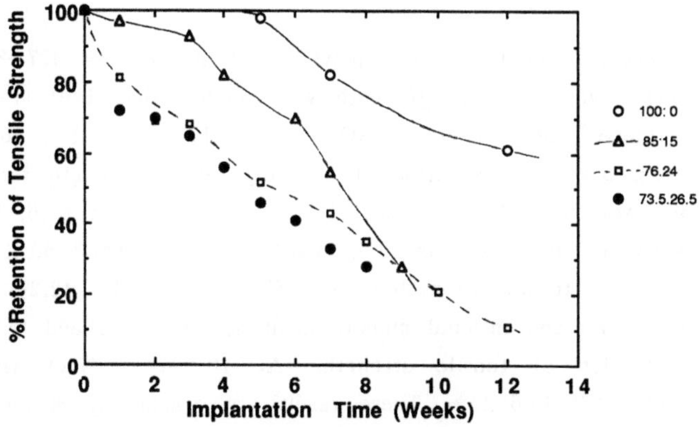

Fig.2 in vitro Degradation of Poly(CL-LA) Suture

A 5-0 monofilament suture was prepared from the following copolymer: a molecular weight $2.2X10^5$ poly(CL-LA) measured by the GPC method. The average diameter of the sutures was 0.165 mm. The sutures were sterilized by ethylene oxide.

After testing the initial mechanical properties, the suture was cut to a length of 10 cm. Then 5 pieces of these sutures were bundled with 5-0 Prolene® sutures, and then implanted in 8 male Japanese white rabbits (Kitayama Labes Co., Kyoto, JAPAN) weighing 2.5-3.0 kg, in their backs, subcutaneously. As for controls, 5-0 PDS®[2], 5-0 MAXON ®[3], 5-0 Prolene ® samples were implanted in the same manner. In addition, for histological evaluation, one suture stretch was made for each sample into the back muscle, and both ends of this suture were ligated tightly on the fascicle of the back muscle. After 1, 2, 4, 6, 8, 10,16 weeks or one year, animals were sacrificed by administration of an overdose of sodium pentobarbital intravenously, to examine the degradation of the sutures and the reaction of the tissues. Bandles of sutures subcutaneously implanted were picked out carefully and straight pull tensile strength tests were done immediately using an AUTOGRAPH®(Shimazu Co. Kyoto, Japan). The span width was 50 mm. The cross-head speed was 20 mm/min. En bloc resections of streched suture with the back muscle were made for histological evaluation.

All the operation and sacrifices were performed in accordance with the Animal Welfare Regulation Guidelines of the Kyoto University Animal Experimentation Committee (1989).

Results:

Initial tensile strength of the poly(CL-LA) suture was 807 ±89 gf, knot-pull strength was 600 ±60.1 gf; both were higher than USP standard. As for the Prolene®: 881±15 gf. MAXON®: 1456± 132 gf. PDS®: 1050±35 gf. At 2 weeks, average %retention of breaking tensile strength for each of the samples was as follows; Poly(CL-LA): 83.2, MAXON®: 85.6, PDS®: 68.0. At 4 weeks Poly(CL-LA) was 83.6, MAXON®: 35.8, PDS®: 50.1,. At 8 weeks %retention of tensile strength of Poly(CL-LA) was 38.2 ±12.2%. At 8 weeks the other two conventional monofilament sutures, PDS®and MAXON,® showed no retention of tensile strength. At 10 weeks %retention of Poly(CL-LA) was still 15.6 ±15%. These result are graphically displayed in Figure 3.

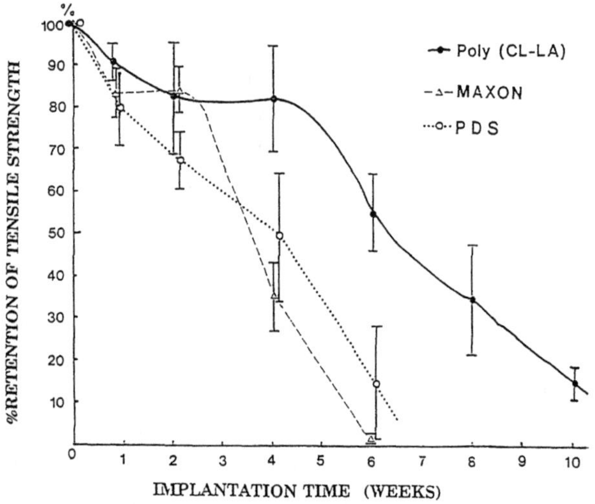

Fig. 3 ***in vivo*** **%Retention of Bioabsorbable Monofilament Sutures**

At 16 weeks the material was surrounded by a whitish thin capsule. Morphological examination using SEM showed that at 8 weeks no morphorogical change was observed on the Poly(CL-LA) suture surface, PDS® and MAXON®had obtained, in the transverse direction, a lot of cracks in the fibers. At 16 weeks no crack on the surface of Poly(CL-LA) was observed (Figure 4). MAXON®and PDS® fibers were macroscopically absorbed at 16 weeks and only the Prolene®sutures, which were used for bundling were recognized. Histological examination revealed that the reaction of this Poly(CL-LA) was minimal. During the acute inflammatory reaction just after the implantation, reaction cells were observed around every suture. But the reaction restricted around the sutures. As shown in Figure 5, in reactive cells, eosinophils were evident around the Poly(CL-LA) sutures. At one year the poly(CL-LA) fibers were replaced with collagen deposition (Figure 6). In close examination of all the preparats, the debris of the materials was seen in some places and foreign body giant cell reaction were observed there. At one year, around the Prolene® suture, connective tissue formation with osteosynthetic reaction was observed and bone formation was seen (Figure 7).

Discussion:

Synthetic bioabsorbable sutures were developed in the 1960's, and because of their good biocompatibility for tissues, they are now widely used in tracheobrochial surgery, as well as general surgery. They are multifilament type sutures. The multifilament (braided) sutures have good handleability, and it is easy to make knots. But for continuous suturing, braided sutures whose surface is not smooth, are not applicable. Because they do not pass through the tissues smoothly, when they are used for continuous suturing, needle holes become large and the tissue is cut by the sutures. Monofilament sutures have smooth surfaces and they are adequate for continuous suturing. As for the monofilament suture, Poly-L-lactide is too stiff and inflexible. Since the glass transition temperature of Poly-L-Lactide is 60 ^{o}C, at room temperature it is difficult to make knots with monofilament fibers of poly-L-lactide and Polyglycolide. Copolymers of L-lactide and ε-caprolactone are bioabsorbable elastic materials and their clinical applications have been studied.[4] As the composition ratio of the L-lactide increases, the glass transition temperature of P(CL-LA) increases. In this study we developed the 5-0 monofilament bioabsorbable suture from a copolymer with composition rate of 85:15.

The decline in tensile strength of this poly(CL-LA) suture was slower than those of other monofilament absorbable sutures which are now on the market. As for MAXON® and PDS®, 50% of the original tensile strength has been reported to remain at 3 and 4 weeks, respectively. In vascular surgery the appropriate absorption rate of sutures is considered to be different depending on the situation in which they are used, such as the location, mechanical strength at the anastmosis, age and general condition of the patients, etc. In rapidly healing tissues, conventional bioabsorbable sutures provide adequate support.[4,5,6,7] In the cases in which longer retention of tensile strength of suture is required, this poly(CL-LA) suture will excel. In addition, because this poly(CL-LA) suture degrades gradually, and the tissue response is moderate, the constriction of the anastomoses site can be avoided. Especially in the pediatric vascular surgery, this suture is promising and appears to have wide applicability.

Acknowledgements: Mr. Won-Ill Cha graciously cooperated in material evaluation and the SEM part of this study. We gratefully acknowledge the manuscript preparation of Miss Naomi Yoshimura.

Reference:

1. Driscoll GL, Baird PJ, Merkelbach PJ, Smith DH (1982) Synthetic absor bable and nonabsorbable microsutures: a histological comparison, Clinical Reproduction and Fertility, 1:151-156
2. Katz AR, Mukherjee DP, Kaganov AL and Gordon S (1985) A new syn thetic momofilament absorbable suture made from polytrimethylene carbonate, Surg. Gynecol. & Obstet., 161:213-222
3. Ray JA, Doddi N, Regula D, Williams JA and Melveger A (1981) Polydioxanone(PDS), A novel monofilament synthetic absorbable suture, Surg. Gynecol. & Obstet., 153:497-507
4. Nakamura T, Hitomi S, Shimamoto T, Hyon S. -H, Watanabe S, Shimizu Y (1987) Surgical Application of Biodegradable Films prepared from Lactide- ε -Caprolactone Copolymers in: Pizzoferrato A, Machetti PG, Ravaglioli A and Lee AJC (eds) Biomaterials and clinical Applications, 759-764, Elsevier Science Publ., Amsterdam
5. Naoe A, (1989) The role of suture material on healing of vascular anas tomosis, J. Jpn. Thorac. Surg., 90:941-948
6. Steen S, Anderson L, Lowenhielm P, Stridbeck H, Walther B, Holmin T (1984) Comparison between absorbable and non absorbable, monofilament sutures for and-to-end arterial anastomoses in growing pigs, Surgery 95:202-208
7. Tons CH, Armbrecht J, Bircks W (1986) The Use of Synthetic Absor bable Suture Materials (Polyglycolid Acid and Polydioxannone) in the Low Presure Circulatory System of Growing Organism, Thorac. Cardiovasc. Surgeon 34:128-131
8. Aarnio P, Harjula A, Lehtola A, Sariola H, Mattila S (1988) Polydioxanone and polypropylene suture material in free internal mamary artery graft anastomoses, J. Thorac. Cardiovasc. Surg. 96:741-5

*Gunze Co. / Aono-cho Ohtsuka Ayabe-city, Kyoto 623, JAPAN.

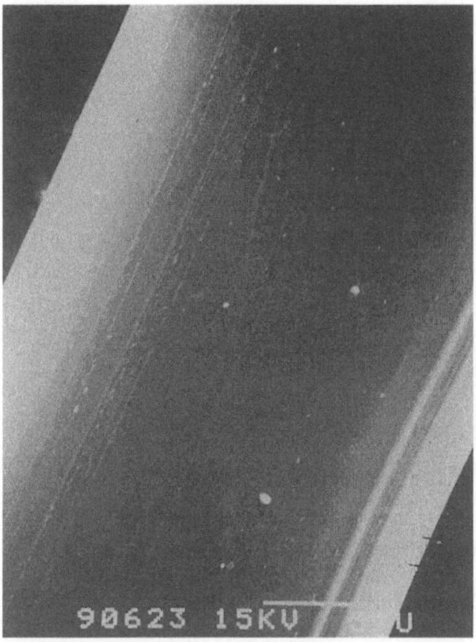

Fig.4 SEM of Poly(CL-LA) at 8 weeks enbeded in the rabbit back muscle.

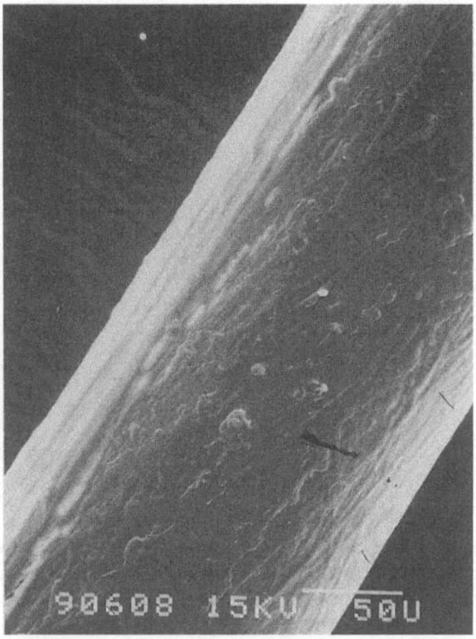

Fig.4 SEM of Poly(CL-LA) at 16 weeks enbeded in the rabbit back muscle.

Fig. 4 SEM of PDS®at 8 weeks, enbeded in the rabbit back muscle.

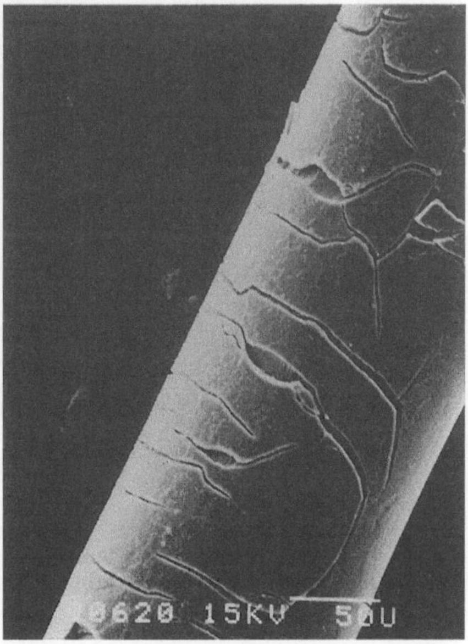

Fig.4 SEM of MAXON®at 8 weeks, enbeded in the rabbit back muscle.

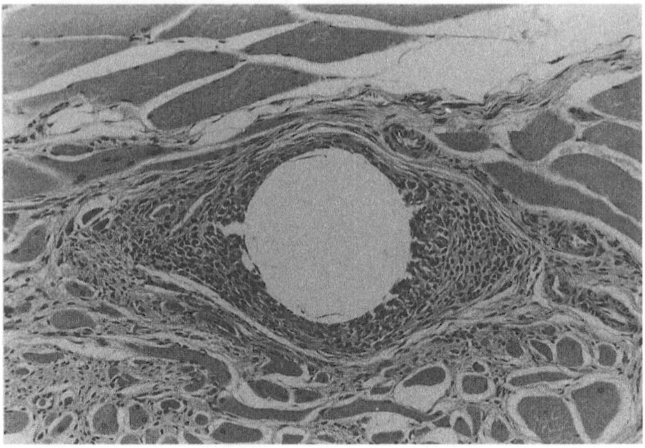

Fig.5-1 Tissue reaction of 5-0 monofilament Poly(CL-LA) at 6 weeks. Hematoxylin and eosin stain, x200. The suture surrounded by irregular zone of inflammatory cells, containing primarily of macrophages with eosinophils.

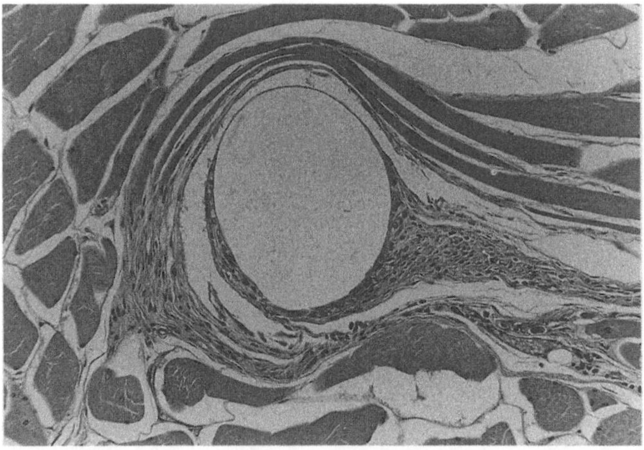

Fig.5-2 Tissue reaction of 5-0 monofilament Prolene®at 6 weeks embeded rabbit back muscle. Hematoxylin and eosin stain, x200. The suture surrounded by a band of fibroblastic cells. The reaction is judged as slight.

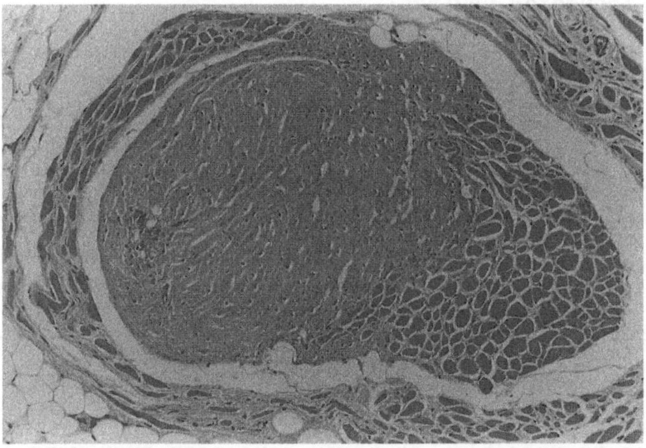

Fig.6 Photomicrograph of 5-0 Poly(CL-LA) implanted in the rabbit at one year, hematoxylin and eosin stain, x200. Absorption of the suture material and their replacement by collagen deposition.

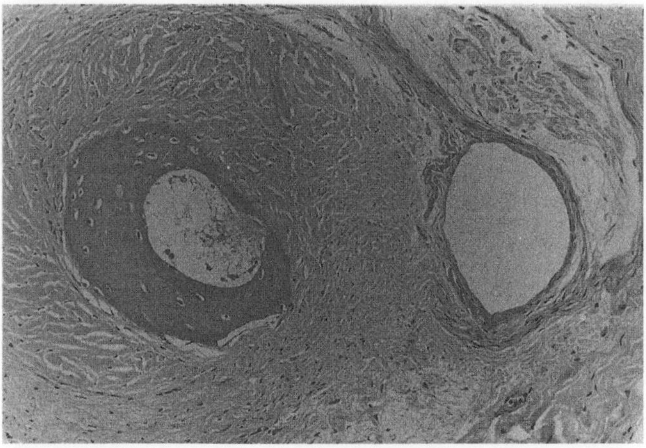

Fig7. Photomicrograph of 5-0 Prolene®implanted in the rabbit at one year, hematoxylin and eosin stain, x200. The connective tissue formation with osteosynthetic reaction (left side) is seen to the left of the remaining suture material (located in the right hole; removed at stainnig).

SELECTED DISCUSSION REMARKS:

Gibbons: Do you know the glass transition temperature of your 85:15-copolymer?

Nakamura: The glass transition temperature is about 42°C.

Thiede: In your histological evaluations you found some fatty cells, as we did in our investigations with absorbable suture materials. Does anyone in the room have any idea, why these fatty cells appear?
For a clinical surgeon, the appearance of these fatty cells is important in eye surgery and also in cutaneous surgery and cosmetic surgery. That is one of the reasons, cosmetic surgeons do not like absorbable materials to fix the skin.

Gibbons: You tend to get fatty cells if you have devitalisation of cells very locally. In my experience, in the fascia lata, this is the usual origin of a fatty residue. So it might be possible, that you have seen just very local cell death as the result of placing a suture there.

Stokes: Near pacemaker electrodes there are always fatty cells surrounding them, that can be traced back to the fact, that the presence of the device prevented the normal motion of the muscle.

Clinical Application of Resorbable Biomaterials in Reconstructive Surgery

K.E. Rehm

Trauma-, Hand- and Reconstructive Surgery
University of Cologne, Joseph-Stelzmann-Str. 9
D - 5000 Cologne, FRG

Introduction

Until now we don't really know how much stability is needed for undisturbed bone healing. Usual metal fixation in fracture treatment is without any doubt a successful procedure. Cortical bone and steel are very different materials, especially concerning the elastic properties. The elasticity constant of bone is only 1/10 of implant steel, and tensile strength is 10 times lower. In the months following internal fixation, bone substance is decreasing as a matter of deviation of force over the implant. The removal of metal implants is followed by a period of weakened bone (stress protection) with the danger of refracture or second fracture. Bone healing is a dynamic procedure, starting with no stability at all up to regular strength. The idea of biodegradable implants depends on the reflected image of the bone, that means a decreasing weightbearing of the material. After months, the whole material has disappeared completely and no secondary surgery is needed. Besides the costs, time of treatment and uncomfort of the patient descends. Polymer materials which are hopeful for this indication are Polyglycolide, Polylactide, Polydioxanone and Polyhydroxybutyrate. The physical data are well known and do not differ fundamentally. They can be improved by higher molecular weight and various kinds of treatment after polymerisation. Copolymers are investigated as well as compound materials. Because we need a complete system for internal fixation, the development of biodegradable implants has five aims:
the replacement material for K-wires, flexible wires, cords, plates and screws.

H. Planck M. Dauner M. Renardy (Eds.)
Degradation Phenomena on Polymeric Biomaterials

At the momentaneous point of view, screws and plates are more difficult to realize. The attainable torque, which causes pressure to the fracture, is very low. At the time, stable fixation of facial fractures is possible, but not of cortical bone at the extremities. The materials and methods where our experience depends on are animal experiments in sheep for ligament replacement and internal fixation as well as clinical experience in ligament augmentation, replacement and support by biodegradable devices. Biodegradable implants seem to have some advantages, compared with metal devices: A second operation for the removal of the implant is not necessary. The stiffness and other biomechanical data can be adapted to the biological aids of the reconstructed tissue. In a dynamic procedure increasing forces are added to the healing tissue. Tissue reactions are very mild, foreign body reactions mostly are missing, especially when the amount of a material is low (5). Our experience concerns applications of cord and tapes, pins and screws.
"The use of biodegradable devices means future treatment of fracture" says P. Rokkanen.

1. Cords and Tapes

When polydioxanone suture material was introduced in Germany by Dociu (1) in 1981, we started to use handbraided cords and tapes for ligament reconstructions.

Within 6 weeks, tensile strength decreased to 58 %. In the bone we saw (7) a close contact of PDS with bony ingrowth (Fig.1), but the material did not induce collagenous tissue inside of the knee joint. With a good coverage of soft tissue, the replaced collateral ligaments in sheep showed sufficient scar tissue along the resected ligaments (4).

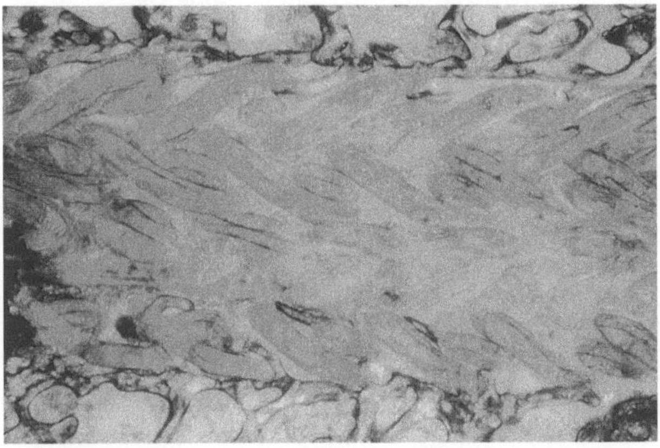

Fig. 1: Polydioxanone cord after 6 weeks of implantation in the cancellous bone of a sheep in combination with ACL-augmentation. Close contact to the trabecular formations.

Polydioxanone cords are able to support ligament replacement and allow early mobilisation. It is also in use for tension band constructions together with pins and rods of compound PLA fibres and matrix (13).

For clinical application, PDS was recommended for ligament augmentation, for securing a ligament suture and as a kind of internal splinting to allow early motion of the extremities after the operation. The following methods were developed and are still in use:

1.1 Acromioclavicular joint separation

The lateral clavicula is connected to the scapula with two systems of ligaments, the strong coracoclavicular and the weaker acromioclavicular ligaments. The suture is able to adapt the torn ligaments, but has to be protected by an implant. But the metal implant must fail by breakage or loosening, because it bridges a still moving joint. We augmentate the sutured ligaments with two loops of PDS cords of two millimeters in diameter, tensile strength is 608 N (2,6) (Fig.2).

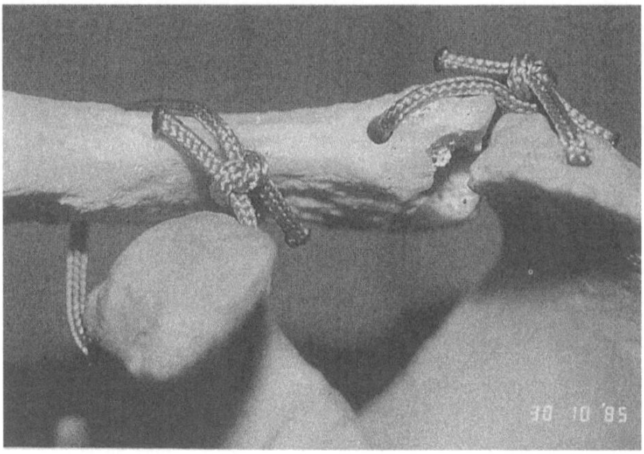

Fig. 2: AC-joint. The PDS-loops in their position. The knots are secured by additional sutures.

1.2 Sternoclavicular joint separation

The medial end of the clavicula is fixed to the sternum and the first rib (Fig.3).

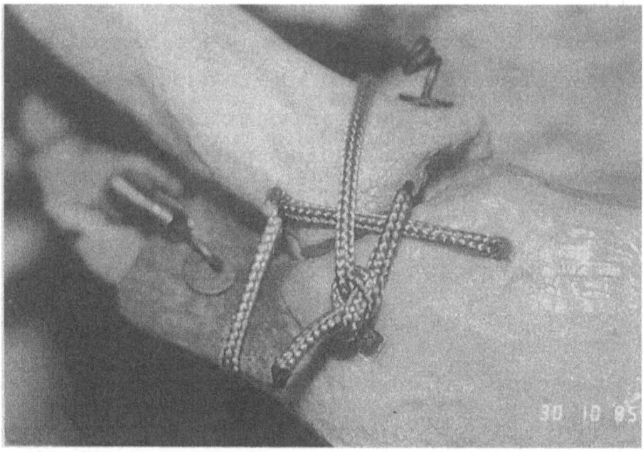

Fig. 3a: SC-joint, fixation of the medial clavicule to the sternal body, one loop around the cartilage of the first rib.

The lesion produces anterior or posterior dislocation of the joint. A lot of methods are described, stiff implants fail. Our original procedure includes two loops, which create a dynamic fixation of the clavicula to the sternum and the first rib.

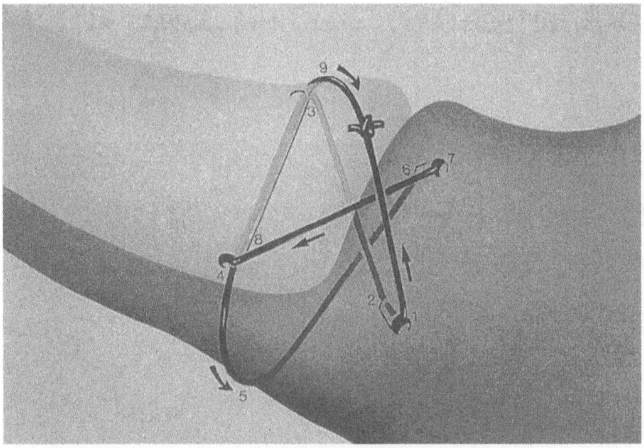

Fig. 3b: Schematic drawing for the position of the bore holes and sequence of the loops.

1.3 Ligamentous instabilities of the pelvis

Ruptures of the symphysis and the sacroiliacal joint can be fixed by strong transosseous sutures with PDS cords (Fig.4,5).

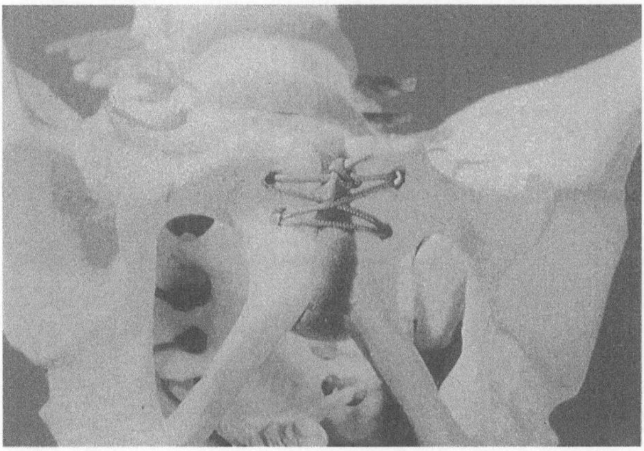

Fig. 4: Symphyseal rupture. Crossed cords at the outside surface of os pubis, inside (not visible) parallel.

The patient is able to turn and move in his bed, weightbearing is not allowed for 6 weeks. Comparable metal constructions hardly can be fixed in the cancellous bone. The danger is high to damage nerves or vessels, when the metal will be removed (8-11).

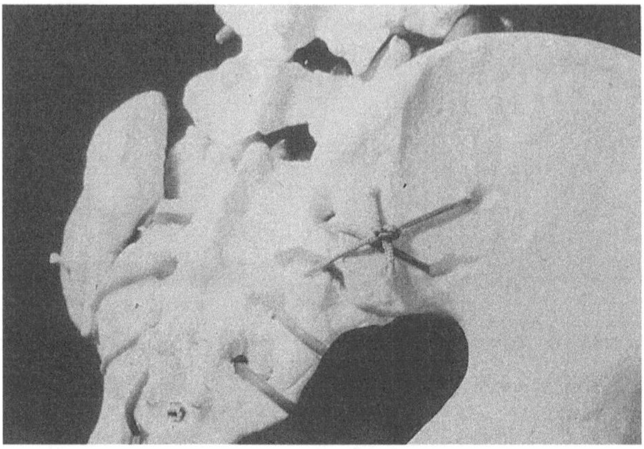

Fig. 5a: Posterior ligamentous instability of the pelvic ring, fixation of the iliosacral joint.

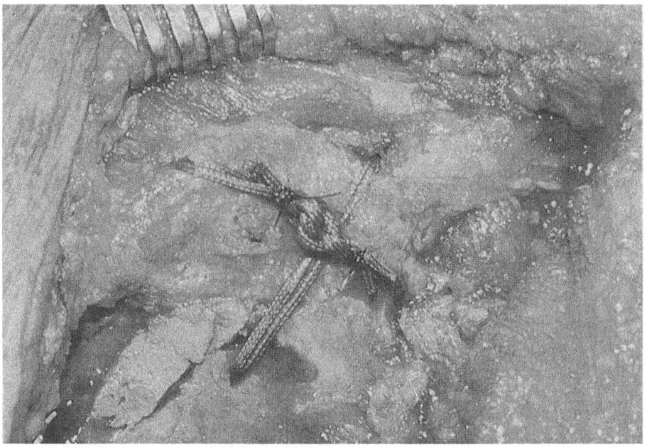

Fig. 5b: A clinical example with additional fractures of iliac bone.

1.4 Anterior cruciate ligament augmentation

Minimal invasive surgery was performed in the replacement of the anterior cruciate ligament by developing a new technique of closed, arthroscopically controlled procedure in 1981 (3,12).

The medial third of the patellar tendon is fixed to a double 2 mm cord (Fig.6).

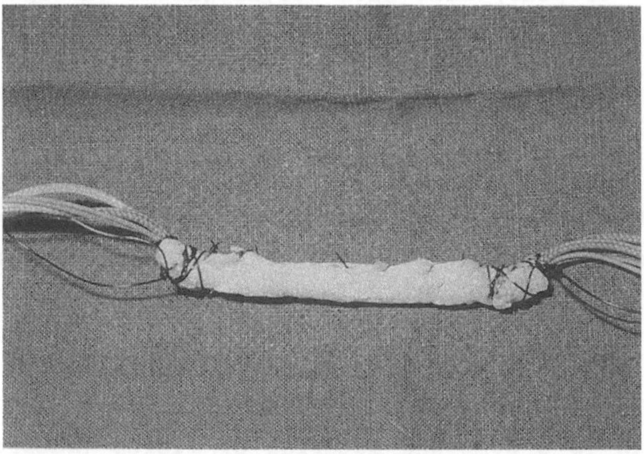

Fig. 6a: Biodegradable augmentation of a third of the patellar tendon. The transplant covers the intraarticular portion of the cord. At both ends strong 1 mm sutures to hold the bony chip.

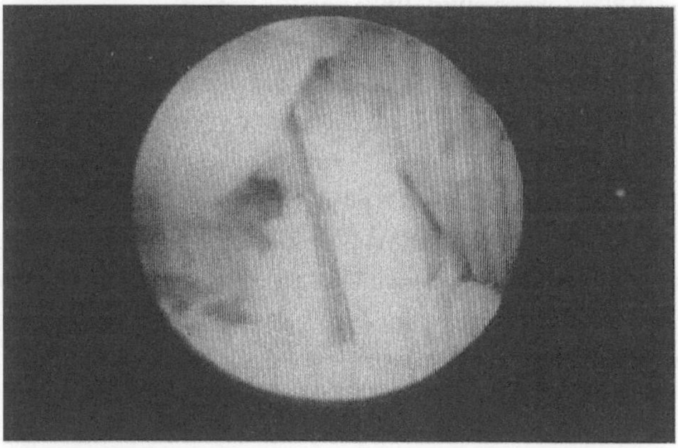

Fig. 6b: Arthroscopy, the tendon is just inserted into the knee.

The bone chips at each side of the autologous transplant is fixed with an additional 1 mm suture. After inserting the transplant into the femoral and tibial bore holes, the cords are fixed outside the joint with screws. Postoperative treatment is possible without any kind of external immobilisation.

2. Pins

For the fixation of small fragments with low load we used various kinds of pins, made with Polydioxanone (Ethipin), Polyglycolide (Biofix) and Polylactide (Polypin).

2.1 Ethipin (Ethicon, Somerville, N.J./USA)

PDS was developed as a monofil, flexible suture material. Shear strength is the lowest of all available pins. It is used in the indication for refixation of small osteochondral fragments with low load.

2.2 Biofix (Bioscience Ltd., Tampere/Finland)

The Biofix pin is a compound material of fibres (sutures) pressed together with a matrix of Polyglycolide. Its strength allows the fixation of larger fragments, but resorption is very fast. Our own experience with malleolar fractures were disappointing. After problems with fistulas we stopped the use of this implant.

2.3 Polypin (Hug, Freiburg-Umkirch/Germany)

After in vitro resorption tests our group with L. Claes chose a copolymer of poly-L/DL-lactide for the Polypin. A new design was made for the surface of the fixation device, contrast application enables visualization of localisation and degradation (Fig.7). A set of instruments makes the pin adaptable for different indications. Good results are possible in radial head fractures and fractures of the apex of various bones like distal radius, metatarsal and metacarpal fractures, some types of ankle transplants (Fig.8,9).

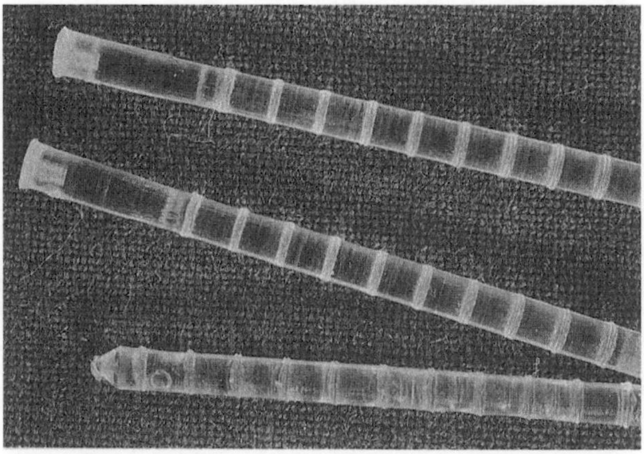

Fig. 7: New biodegradable pin ("Polypin") with X-ray contrast in the head.

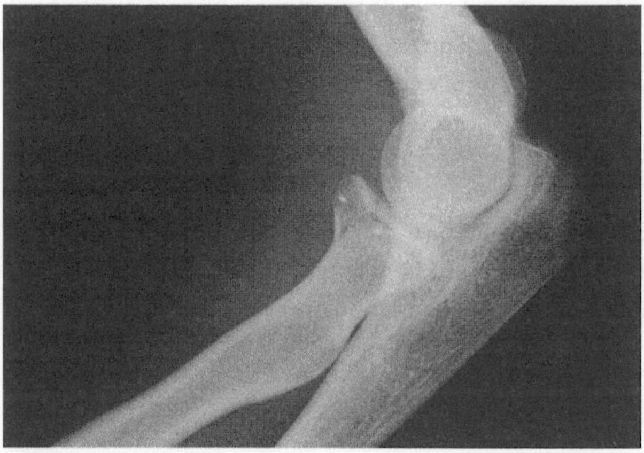

Fig. 8: Radial head fracture Type B 2, open reduction and fixation with Polypin.

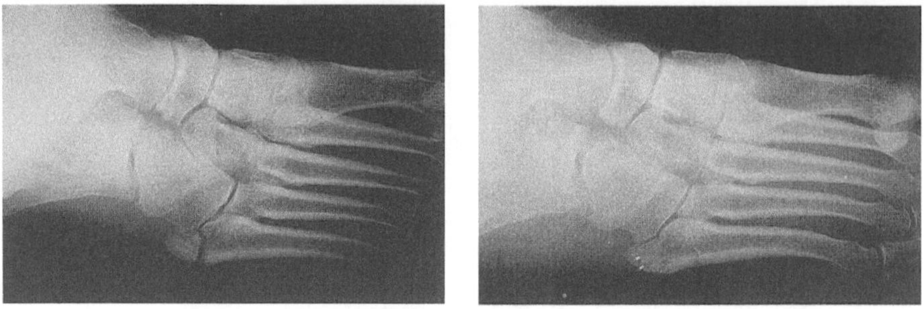

Fig. 9: Pre- and postoperative X-ray of a metacarpal fracture in the technique of tension band wiring.

3. Screws

For screws improved physical properties are needed, so Polylactide is used by all companies and groups. Bioscience, Johnson & Johnson, Physis and also our group are testing threaded implants. Plates are in use for maxillofacial indications, cortical bone fixation is not recommendable at the time.

3.1 Bioscience screws (Bioscience, Tampere/Finland)

The thread follows the usual AO/ASIF design (Fig.10). The fibre reinforced screw is not stiff enough for clinical use in cortical bone. Compression of fragments cannot be realized.

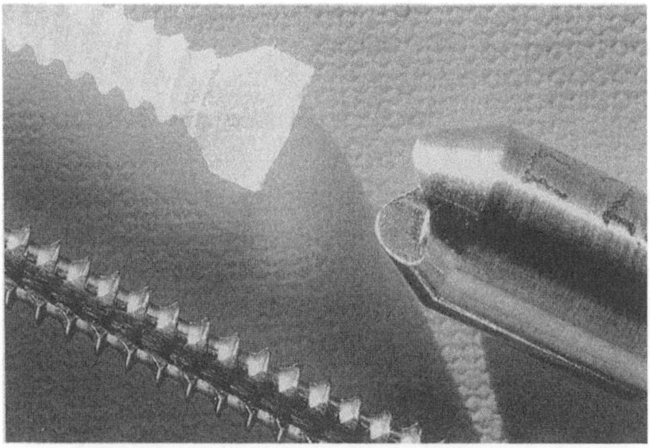

Fig. 10: Bioscience screw with instruments.

3.2 Johnson & Johnson screws (J & J, New Brunswick, N.J./USA)

Johnson & Johnson offers a very stiff PLLA screw exclusively for ankle fractures. High molecular weight and a special treatment of the polymer result in the stiffest polymer screw. The thread is small. Torque limiting screw-drivers allow proper application. We use the 4.0 and 4.5 mm implant in a clinical test in comparison to metal screws (Fig.11).

Fig. 11: PLLA-screw by Johnson & Johnson for internal fixation of a medial ankle fracture.

There is no difference in the results, no complications at a maximum follow up time of two years.

3.3 Polyscrew and Polyplate (Hug, Freiburg-Umkirch/Germany)

Our 2.7 mm screws are tested together with small plates in maxillofacial fractures. In this combination anglestable fixation is possible. Bending and flexion of the plates enable perfect fit to various bone surface formations (Fig.12).

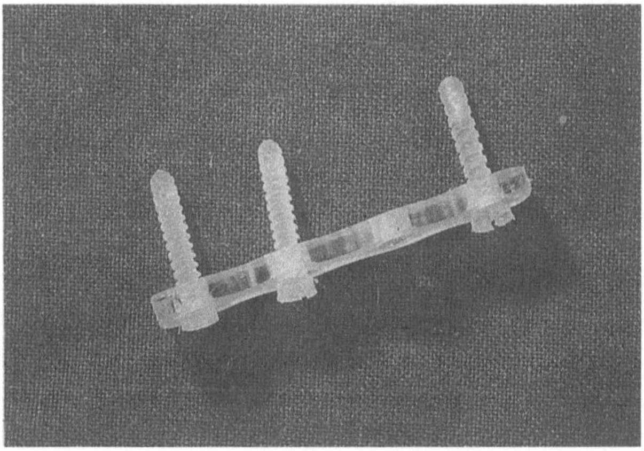

Fig. 12: "Polyplate" and screws 2.7 mm for maxillofascial surgery

4. Other applications

Biodegradable polymers are possible for a lot of indications. A marrow spacer helps to save autologous bone material. A plug for closing the bone marrow is used for endoprosthetic joint replacement. Fibres are used for filling of large bone defects without mechanical load. Biodegradable drug delivery systems are interesting alternatives to expensive drug pumps.

Summary

Bands and cord are used for AC and SC joint separation and ligamentous instabilities of the pelvic ring, pins for osteochondral fractures and ankle fracture fixation. A new pin for the fixa-

tion of small fragments was designed together with L. Claes of the Ulm University. Our Polypin has a new design, new instruments and is visible in X-ray. Dislocation and biodegradation are able to be visualized. Osteochondral fractures, especially of the radial head, are good indications for pins.
A clinical study with Johnson-PLA screws for ankle fractures shows good results and no specific complications.

References

1. Dociu, N., P. Hein:
   PDS - ein neues chirurgisches Nahtmaterial
   Ethicon OP-Forum (1981) 108
2. Keller, H.W., K.E. Rehm:
   Die Versorgung der kompletten Schultereckgelenkssprengung ohne metallisches Implantat
   Unfallchirurg 94 (1991) 511-513
3. Rehm, K.E.:
   Die geschlossene Kreuzbandplastik
   Unfallchirurgie 7 No. 2 (1981) 55-59
4. Rehm, K.E., K.H. Schultheis, P. Bopp, H. Ecke:
   Biomechanische Untersuchungen vom resorbierbaren Bandersatz und deren klinische Bedeutung
   Langenbecks Arch. Chir. Suppl. (1984) 207-211
5. Rehm, K.E., B. Kaletsch:
   Mikrogefässanastomosen mit resorbierbarem oder nicht resorbierbarem Nahtmaterial
   in: Biomaterialien und Nahtmaterial, S. 298-302
   Hrsg.: H. Rettig
   Springer Verlag, Berlin - Heidelberg 1984
6. Rehm, K.E.:
   Versorgung der Schultereckgelenkssprengung ohne metallisches Implantat
   in: Biomechanik der gesunden und kranken Schulter, S. 47-48
   Hrsg. H.J. Refior, W. Plitz, M. Jäger u. M. Hackenbroch
   Thieme Verlag, Stuttgart - New York 1985
7. Rehm, K.E., K.H. Schultheis:
   Bandersatz mit Polydioxanon (PDS)
   Unfallchirurgie 5 (1985) 264-273
8. Rehm, K.E.:
   Biodegradable implants in Traumatology and Reconstructive Surgery: Stage of development and clinical significance
   Polymery W Medycynie
   PWN-Polish Scientific Publishers T XVIII No. 3 (1988) 117
9. Rehm, K.E.:
   Entwicklungsstand und klinische Bedeutung von resorbierbaren Osteosynthesematerialien
   Hefte zur Unfallheilkunde 200 (1988) 663

10. Rehm, K.E., H.J. Helling, L. Claes:
Biologisch abbaubare Osteosynthesematerialien
in: Jahrbuch der Chirurgie (1989) S. 223-232
Biermann Verlag, Zülpich 1989
11. Rehm, K.E.:
Biodegradable Implants in Traumatology and Reconstructive Surgery
Chemia Stosowana 34, No. 1 - 2 (1990), 41-48
12. Rehm, K.E.:
Kreuzbandaugmentationsplastik mit Patellarsehne und PDS-Kordel
Serie "Im Dienste der Chirurgie" 72
Ethicon, Hamburg (1990)
13. Zieren, H.U., W.A. Holzmüller, J. Rosenberger, K.E. Rehm
Sind Patellazuggurtungen mit resorbierbaren Materialien möglich?
Unfallchirurg 94 (1991) 634-639

BIODEGRADABLE PLA VERSUS STAINLESS STEEL INTRAMEDULLARY DEVICES FOR FRACTURE FIXATION. A COMPERATIVE HISTOLOGICAL STUDY

M. van der Elst, A.R.A. Dijkema, C.P.A.T. Klein, P. Patka, H.J.Th.M. Haarman

Free University Hospital, Department of Traumatology

P.O. Box 7057, 1007 MB Amsterdam, The Netherlands

INTRODUCTION

Throughout history man has possessed a need to repair or replace parts of the human body that have become none functional. For hundreds of years crutches and wooden legs have been examples of simple yet meaningfull atttempts to improve body functions.

As man's knowledge and technical skill increased, so did his ability to provide more sophisticated surgical procedures and implants designed to aid the natural process of self-repair.

Most implants currently used in surgery are of stainless steel origin, due to their satisfactory mechanical properties and availability.

Fortunately and unfortunately the breaking strenght of stainless steel is about ten times the breaking strenght of human cortical bone.

Steel is practically non-reactive and can be shaped according to the dimensions of the fractured bone. Stainless steel is relatively inexpensive and mechanically predictable.

However significant problems are associated with stainless implants for fracture fixation, including:

1. The high modulus of elasticity of metal implants as compared to bone which results in the implant retaining a large fraction of the mechanical load applied to the bone even after the fracture has healed. This phenomenon is known as "stress shielding".
2. As stainless steel implant poses an increased risk of infection, even after an umcomplicated postoperative course.

H. Planck M. Dauner M. Renardy (Eds.)
Degradation Phenomena on Polymeric Biomaterials

3. A second operation to remove the implants after the fracture has healed is ofter necessary and increases Health Care costs.

A material that would slowly be absorbed by means of hydroloysis or enzymatic degradation would obviously be advantageous. A second operation would not be necessary and stress-shielding would not occur, resulting in a better product from boneremodelling.

CHOICE OF MATERIAL

In 1971, Kulkarni first described biodegradable polymeres for fracture fixation. Since then, a large variety of materials have been tested. The mechanical properties of the polymeres used show a considerable variation. The distribution of molecular weight plays an important role. The cristalllinity, the amount of free monomeres, and the stereotactic configuration are all factors determining the mechanical properties. The ultimate strenght of poly-lactic acid rods resembles the ultimate strenght of the cortical bone, that normally varies from about 10 up to 150 megapascals. Mechanical properties of biodegradable implants can be fine-tuned using composites. The basic material for production of implants can be manufactured in a simple manner. Polylactic acid with a high mean molecular weight, a large degree of cristallinity, and a small amount of free monomeres is a promising material for further study.

In an atttempt to develop a biodegradable intramedullary device for fracture fixation the tissue reaction induced by stainless steel in comparison to polylactic acid interlocking nails were studied in an animal model.

AIMS OF THE STUDY

1. The mechanical properties of the PLA rod before and after implantation.
2. The chemical properties of the PLA rod used before and after implantation.
3. The tissue reaction from materials used.

MATERIALS AND METHODS

A combined in vitro and vivo experience was performed. The mechanical properties is the polymer which was used in this experiment must approach the mechanical properties of an ideal implant material. Therefore, we had to use an implant of considerable dimensions to provide adequate strenght. We this limitation in mind, an experimental model was developed in which the animals used provided us with medullary cavities of acceptable diameters for implantation. The diameter of a porchine femur matches the diameter of a humam femoral bone. The implantation of the interlocking nail is relatively simple and postoperatively the bone is well covered by large musclegroup that promote quick healing. In view of the fact that pigs are fast growing animals and difficult to handle when they are adult, we conducted the experiments in young animals. Considering the fact that resorption rate depends on the implant's form, we used implants that resemble the implants used in humans. This implants were prepared from a large piece of polylactic acid. The polylactic acid rod diameter is 12 mm and the length is 10 centimeters. They were locked proximally and distally of midshaft femur osteotomie using transverse placed polylactic acid rods. In the controlgroup the osteotomised femur was stabilized using a hollow steel implant with identical dimensions, locked with stainless steel screws. The left femur of twenty female pig, aged three months, weigth 35 kg, were operated om while the animals were under general anaesthesia. The femur was reamed using a flexible drill. A midshaft osteotomy was performed with an oscillating saw. The rods were introduced into the medullary canal and locked proximally and distally. X-rays were taken one week after implantation to check the osteosynthesis. Ten pigs were terminated after four weeks, and the other ten were terminated after twelve weeks. Postmortum, the femurs and implants were examined histologically, mechanically and chemically. The mechanical tests were performed with a three point bending test.

The mechanical properties of the PLA interlocking nails used, showed a considerable decrease in bending strength after one and three months. Chemical analysis of the rods before and after implantation showed a considerable decrease in mean molecular weigth. The fracture healing was also studied histologically four, and twelve weeks after implantation. The distal part of the fractured femur, including a part of the interlocking nail was used to study the bone-implant interface. The remaining proximal part of the interlocking nail was removed from the femoral shaft. After mechanical testing of the femur and rod, the proximal and mid-callus parts were studied histologically looking specifically after the new bone formation. The samples were prepared and embedded according standard PMMA embedding procedures.

Ten to twenty micron thin sections were cut from the undecalcified bone samples embedded in PMMA using especially prepared saw microtome.

Methylene blue, fuchsine and Masson tri-chrome staining was used to show the degree of mineralisation of the bone. Histological evaluation of the sections was performed using convential and polarised light microscopy. The goal of the histological examination was to determine:

1. The presence or absence of a fibrous connective tissue layer around the implants.
2. The presence or absence of macrophages.
3. The presence or absence of osteoblasts and osteoclasts.
4. Evidence or signs of an inflammatory respons.

RESULTS

Examining the material obtained from animals with a stainless steel implant, we saw a circular fibrous connective tissuelayer around the implant. This fibrous tissue layer contained fibroblast and microphages. Around the fibrous tissuelayer we saw bonetissue in different stages an maturity containing osteoclasts and osteoblasts. There was no sign of a serious inflammatory respons.

The material obtained from animals with the PLA implant also showed a circular fibrous connective tissue layer around the implant. This fibrous connective tissue layer Also contained fibroblasts and microphages. Around the fibrous tissuelayer we saw bone in different stages of maturity. Osteoclasts and osteoblasts were present. Neither in one month specimens of three month specimens there was evidence of an inflammatory response. Microradiograms showed no difference in bone structure between the two groups. We also performed image-analysing techniques on our slides. No significant difference was found between the groups in respect of the amount of bone tissue per surface area. The results from our experiments bring us to the following conclusions:

1. On a macroscopic level, a bit more callus was seen in the polylactic acid group. This effect is possibly due to the difference in mechanical properties of the materials used. Especially the difference in stiffness. The polylactic acid rods allowing the fracture to bear more of the load than the stainless steel rods.
2. Histological evaluation show that the polylactic acid and stainless steel devices appear to induce similar tissue reactions.

REFERENCES

1 Gogolewski S (1990). Resorbable polymers for internal fixation. Davos

2 Vainionpaa S, Vithonen K, Mero M, Rokkanen P, Kilpikari J, Tormala P (1986) In: Arch. Orthop. Trauma. Surg. 106:1-4. Fixation of experimental osteotomies of the distal femur of rabbits with biodegradable material.

3 Daniels AU, Chang MKO, Andriano KP (1990), In: Journal of Applied Biomaterials, Vol. 1, 57-58. Mechanical pro pertiies of biodegradable polymers and composites proposed for internal fixation of bone.

4 Tonino AJ, Davidson CL, Klopper PJ, Linclau LA (1976) In: Journal of Bone and Joint Surgery, Vol. 58-B. Pro tection from stress in bone and its effects.

5 Nakamura T, Hitomi S, Watanabe S, Shimizu Y, Jamshidi K, Hyon S-H, Ikada Y (1989) In: Journal of Biomedical Materials Research, Vol. 23, 1115-1130. Bioabsorption of polylactides with different molecular properties.

6 Williams DF (1976), Biocompatibility of clinical implant materials. Vol. 2, Florida.

7 Leenslag JW, Gogolewski S, Pennings AJ (1984): In: Journal of Applied Polymer Science, Vol. 29, 2829-2842. Resorbable Materials of Poly(l-lactide). Influence of secondary structure on the mechanical properties and hydrolyzability of Poly(l-lactide) fibers produced by a dry-spinning method.

8 Hollinger JO, Battistone GC (1976) In: Journal of Bone and Joint Surgery, No. 207. Biodegradable bone repair materials.

9 Eitenmuller J, Gerlach KL, Schmickal T, Muhr G (1987) In: Chirurg, 58, 12. Semirigide Plattenosteosynthesen unter Verwendung absorbierbarer Polymere als temporare Implantate.

10 Ewers R, Forster H (1985) In: Dtsch Z Mund Kiefer Gesichts Chir 9, 196-201. Resorbierbare Osteosynthesematerialien.

11 Kulkarni RK, Moore EG, Hegyeli AF, Leonard F (1971) In: Journal of Biomedical Material Research, Vol. 5, 169--181. Biodegradable Poly(lactic acid) polymers.

12 Christel PS, Vert M, Chabot F, Abols Y, Leray JL (1983) In: Biomaterials and Biomechanics. Polylactic acid for intramedullary plugging.

13 Vainionpaa S (1987) Biodegradation and fixation properties of biodegradable implants in bone tissue. Dissertatie, Helsinki.

14 Bostmann O, Hirvensalo E, Vainionpaa S, Vihtonen K, Tormala P, Rokkanen P (1990) In: International Orthopaedics, 14:1-8. Degradable polyglycolide rods for the internal fixation of displaced bimalleolar fractures.

15 Perren SM (1979) In: Clinical Orthopaedics and Related Research, No. 138. Physical and biological aspects of fracture healing with special reference to internal fixation.

SELECTED DISCUSSION REMARKS:

Rehm: Do you think the observation time of 3 months is enough with polylactide?

van der Elst: No, of course it is too short. But we wanted to develop an experimental model to see whether it could be implanted and whether the animal was correct for that.

Rehm: The tissue reaction will happen much later when the material is completely dissolved. When you use polylactide, it is about 2 years.

van der Elst: Yes, probably. And what struck me with the talk of Prof. Williams is, that you also have a different activity of, for instance, macrophages, even in between weeks. So maybe it is not only too short, but also you should do it more frequent.

Rehm: Did you perform osteotomy and did you have problems with non-unions?

van der Elst: We performed complete osteotomy and we had no problems with non-unions. But one point that should be considered here is, that we used young healthy pigs, and that the population, we are talking about in humans, is old and osteoporotic.

Rehm: There are some experiments made by a Finnish group in rabbits and they had also good results, but they had a large number of non-unions. And we tried it also with rabbits and dogs and had no good results with the intramedullar implants, so that we at the time cannot recommend to try it in the clinical application.

Rozema: You talked about extensive callus formation. In which areas particularly can you see that?

van der Elst: There was an extensive callus growth at the site of the fracture. Of course the nail is inside the medulla, so you don't have too much callus formation there. And there was a little bit more callus in the PLA-group, but it was not significant in such a little number of animals. But we think that it is because the PLA is less stiff, it gives more possibility for movement and thus it enhances the callus formation.

Rehm: I think that it is one of the problems of the study that the animals were too young. Usually we make the experiments with adult animals, with sheeps or dogs, and the callus formation differs in this case. There is one experiment you can do in young animals. If you test the reaction of the growth zone, the epiphysial zone, it is allowed to use this material and it is very interesting. Pertti Rokkanen made some experiments and he saw, that the degradation at the epiphysial plate is much faster and the reconstruction at the epiphysial plate is better. But in the fracture of long cortical bone, you should use adult animals.

van der Elst: I agree.

ROUND TABLE DISCUSSION AT THE END OF THE CONFERENCE

Chairman: Prof. Dr. G. Egbers
Participants: Dr. D. Gibbons
Dr. H. Planck
Prof. Dr. K. Rehm
Prof. Dr. R. Guidoin

Egbers: The conclusions I draw from this conference are: There are indeed very good results using resorbable biopolymers and there are many new applications. But on the other hand there are still many effects with regard to resorption and degradation, that are not fully understood. Concerning the testing we have to say, that animal tests are worldwide under criticism and the importance of in vitro simulation grows. And there is a big need for standardization, also according to the end-use of the implant. Because you can get almost any result depending on the site or the species.

Rehm: I take the standpoint of the practical user, the surgeon, who has the duty to argue for the security of his patient. In the orthopedic field only very few resorbable materials are accepted for the clinical use. At the time, FDA admission exists only for the Ethipin and the Biofix. Cords and tapes are also, like suture materials, accepted in Europe, but not in the States. So the clinical cases I showed you, are at the level of the clinical experiment. A lot of polymeric implants showed problems, so at the time, in orthopedic surgery metal-implants are the regular implants and the others are the experiments. We intend to replace flexible wire by bands, that is possible in some applications.
We intend to replace the stiff wire by pins, that is also possible in small fragments with minimum load. And we can replace some screw indications by degradable screws at the time, but in this case no load is admitted to the implant. Intramedullary devices we cannot recommend, because in the

animal experiments the results are insecure and the complication rate is very high. For testing of new implants the test procedure should be the following: After the experimental investigation and if that is positive, you have to start a pilot study with only a small number of patients, and after that a multicenter study. And after 5 years you can go to the market.

Guidoin: I do believe, that in vitro investigations are of permanent importance to evaluate the functionality and the durability of a material. And the possibility to investigate toxicology, because it is quite important to know, that a product is non-toxic, non-mutagenic and non-carcinogenic. And the possible degradation products should also be safe. And this is something we should pay much more attention to.

Investigation in vivo is of permanent importance and should be part of the test programme. The problem is, that animal experimentation is limited, because usually you are working with healthy animals, because no council would allow you to do work with diseased animals. The last step is the clinical evaluation. Only in patients you see the longterm effects and the complications. And we should try to obtain as many materials as possible at reoperations, not only at complications with the prosthesis, and to have as many autopsies as possible, particularly when the patient died due to causes totally unrelated to the prosthesis or implant.

Gibbons: One comment concerning animal experiments is: We cannot avoid animal use, but in some cases we may have to admit, that there isn't an adequate animal model. The animals are different in mechanics and in many other aspects. In blood, for example, each species is slightly different in its fibrinolytic activity and its activity of platelets. So you have to consider these differences, when you choose an experimental animal.

Planck: We need definitions for in vitro testing to predict the behaviour of a product, not only a material, in vivo. And we need a correlation of in vitro tests to in vivo behaviour and that is very difficult. To simulate long-term observations, we need accelerated tests, which is a big problem. As we have discussed during the conference, especially accelerating test conditions can change the chemical structure of a plastic material, and so correlation of in vitro and in vivo data is difficult.

Brandwood: As Dr. Gibbons mentioned, many polymers contain antioxidants to prevent degradation. We prepared a polyurethane, which had the same formulation as Pellethane, but we left out all the antioxidants. And we wonder, what those antioxidants in the commercial materials are doing, because the behaviour of the material we prepared without any of the additives, was precisely the same as that of Pellethane. Both, in accelerated tests and in vivo, we saw the same stress cracking, the same degradation.
My second question is: Can someone give a comment which is the best implant site for a degradation experiment for blood contact materials?

Guidoin: What we have to realize is that the main cause for degradation in blood is the lipid uptake. For example we had to abandon the silicone membrane oxygenators because of lipid uptake. After a few hours the membrane did no longer transfer CO_2 and oxygene as it should have done. Polyurethane is another material that fails because of lipid uptake and also bilirubin uptake. In vitro, it is possible to expose materials to lipid solutions. But at the present time, we have no experimental model to mimic the lipid uptake, because, particularly in dog, the blood properties are very different. And for that reason, it is so important to look at explanted materials and to investigate the course of failures.

Tyler: I want to comment again on the subject of antioxidants: From some of the work we have done with SIMS-studies on polyurethanes with antioxidants, we know, that in an air environment or in a vacuum environment, that is very hydrophobic, the antioxidants tend to concentrate on the surface. We have not yet looked at hydrated studies with the antioxidants, but we have looked at hydrated studies with a number of low molecular weight hydrophobic molecules. And when you hydrate these polymers and put them in an aqueous environment, those materials, that so much like to be on the surface in the air and the vacuum environment, migrate into the bulk of the polymer, because they don't want to be in the water. And I think, it is a very reasonable hypothesis, that the antioxidants that are put in the polymers for atmospheric use and other uses with the surface in air, end up in the bulk of the polymers, when the surface is unprotected. And I think, that is something, that perhaps should be investigated, that the antioxidants, which are used for polyurethanes in air, may not be appropriate for polyurethanes in the aqueous environment.

Planck: In my opinion, this is why we should make tests with the final product in the final stage.

Rozema: I have a question to Prof. Rehm:
You suggested a period of 5 years of surveillance of the patient. Would that be enough?

Rehm: When we test a material that degrades in 1 year, I think, that is enough for biocompatibility and tissue reaction. But we cannot get the tissue, because we have no reason to reoperate the patients. So this is a question we have to ask the animal experiment. And I would like to agree to Dr. Planck: I feel it is very important to test the material at the end of the development, the final product in such conditions, in which the surgeon uses it. Because the manufacturing of the material and the sterilization process and all this changes the material. And we have wrong information, when we use the primary material as the factory delivers it.

Rozema: We have used the as-polymerized PLLA for 5 years now in humans. It is a very slowly degrading material and up to 4 years we see a sort of renewed reactivity of the tissue response. One of the qualities of a slowly degrading material could be, that it has a very moderate tissue response, maybe that is one of the clues of the success of the material. But in the end there is a phase which could give clinical complications. And therefore I think you have to make also clinical trials to be sure, if the material is on the whole resorbed without for example remnants in the lymph nodes. So 5 years are a very optimistic view.

Rehm: A similar example is the Biofix rod. It disappears within 8 weeks, and 2 years after the implantation we have the reaction in the bone.

Guidoin: I would say, 5 years seem to be very long for any investigation and analysis. But you have to realize, that a team of scientists is probably cheaper than a team of lawyers. And we have to be very realistic, in future there will be a lot of court actions, because the organisations of consumers are getting more and more active. So this is something we should have in mind and when we develop a new material or new prosthesis, we should rely on very firm data from both: in vitro and in vivo investigations, and clinical investigations as well. Because a judge will not accept a nice letter of any professor as a guarantee.

van den Berg: Professor Rehm, do you also have a suggestion for the number of clinical tests?

Rehm: I can give you just my feeling, not exact dates. I think, you should make a pilot study with 40 or 50 cases. And the multicenter study should bring about 50 patients per clinic, that gives in total about 500 patients.

Gibbons: With all the variables from site to site, surgeon to surgeon, material to material, patient to patient, to get good statistics you need something like 1500 patients over more than 3 years.

van der Elst: What do you think of simulating these time-consuming long-term studies by putting depolymerized substances for instance in animals?

Gibbons: I think, for simulating a pin, you can make a cylindrical hole by a drill, put in the particular volume of low molecular weight substance and look at the responses, when the body is suddenly faced with 2 ml of fluid containing a high concentration of low molecular weight substance. I think, it is an acceptable way, if you choose the right implantation site.

KEYWORDS

General keywords: Degradation, Polymers, Biomaterials, Resorption

The page numbers are referring to the first page of the paper concerned.

AUTHOR INDEX

Ascherl, R., Orthopädische Klinik und Poliklinik rechts der Isar, TU München, Ismaninger Str. 22, D-8000 München 80

Blümel, G., Institut für Experimentelle Chirurgie der TU München, Ismaninger Str. 22, D-8000 München 80

Boering, G., Dept. of Oral and Maxillofacial Surgery, University Hospital Groningen, Oostersingel 59, NL-9713 EZ Groningen

Bos, R.R.M., Dept. of Oral and Maxillofacial Surgery, University Hospital Groningen, Oostersingel 59, NL-9713 EZ Groningen

Brandwood, A., Centre for Biomedical Engineering, University of New South Wales, Kensington, NSW, Australia 2033

Buchholz, B., Boehringer Ingelheim KG, Entwicklung Chemikalien, D-6507 Ingelheim

Chatelier, R.C., CSIRO Division of Chemicals and Polymers, Private Bag 10, Clayton, Vic, Australia 3168

Dauner, M., Institut für Textil- und Verfahrenstechnik , Körschtalstr. 26, D-7306 Denkendorf

de Bruijn, W.C., AEM Unit, Clinical Pathological Institute I, Erasmus University of Rotterdam

Dijkema, A.R.A., Free University Hospital, Department of Traumatology, P.O. Box 7057, NL-1007 MB Amsterdam

Früh, H.-J., Institut für Experimentelle Chirurgie der TU München, Ismaninger Str. 22, D-8000 München 80

Gibbons, D.F., 3M Center, Biosciences Laboratory, St. Paul, MN 55144-1000, USA

Glasmacher-Seiler, B., Helmholtz-Institute for Biomedical Engineering, Pauwelstr. 30, D-5100 Aachen

Gunnatillake, P.A., CSIRO Division of Chemicals and Polymers, Private Bag 10, Clayton, Vic, Australia 3168

Haarman, H.J.T.M., Free University Hospital, Department of Traumatology, P.O. Box 7057, NL-1007 MB Amsterdam

Hyon, S.H., Research Center for Biomedical Engineering, Kyoto University, 606 Kyoto, Sakyo-Ku, Japan

Jansen, H.W.B., Dept. of Histology and Cell Biology, University of Groningen, NL-9713 EZ Groningen

Jonas, J., Helmholtz-Institute for Biomedical Engineering, Pauwelstr. 30, D-5100 Aachen

Klein, C.P.A.T., Free University Hospital, Department of Traumatology, P.O. Box 7057, NL-1007 MB Amsterdam

Lünstedt, B., Chirurgische Universitätsklinik Würzburg, Josef-Schneider-Str. 2, D-8700 Würzburg

Matsui, T., Research Center for Biomedical Engineering, Kyoto University, 606 Kyoto, Sakyo-Ku, Japan

Mau, H., Institut für Experimentelle Chirurgie der TU München, Ismaninger Str. 22, D-8000 München 80

McCarthy, S.J., CSIRO, Division of Chemicals and Polymers, Private Bag 10, Clayton, Vic, Australia 3168

Meijs, G.F., CSIRO Division of Chemicals and Polymers, Private Bag 10, Clayton, Vic, Australia 3168

Müller, E., Institut für Textil- und Verfahrenstechnik, Körschtalstr. 26, D-7306 Denkendorf

Nakamura, T., Research Center for Biomedical Engineering, Kyoto University, 606 Kyoto, Sakyo-Ku, Japan

Nijenhuis, A.J., Dept. of Polymer Chemistry, University of Groningen, NL-9713 EZ Groningen

Nishiya, K., Gunze Co., Aono-cho Ohtsuka Ayabe-city, Kyoto 623, Japan

Noble, K.R., Centre for Biomedical Engineering, University of New South Wales, Kensington, NSW, Australia 2033

Okumura, N., Research Center for Biomedical Engineering, Kyoto University, 606 Kyoto, Sakyo-Ku, Japan

Patka, P., Free University Hospital, Department of Traumatology, P.O. Box 7057, NL-1007 MB Amsterdam

Pennings, A.J., Dept. of Polymer Chemistry, University of Groningen, NL-9713 EZ Groningen

Planck, H., Institut für Textil- und Verfahrenstechnik, Körschtalstr. 26, D-7306 Denkendorf

Ratner, B.D., Dept. of Chemical Engineering and Center for Bioengineering, BF-10, University of Washington, Seattle, WA 98195, USA

Rau, G., Helmholtz-Institute for Biomedical Engineering, Pauwelstr. 30, D-5100 Aachen

Rehm K.E., Trauma-, Hand- and Reconstructive Surgery, University of Cologne, Joseph-Stelzmann-Str. 9, D-5000 Cologne

Reul, H., Helmholtz-Institute for Biomedical Engineering, Pauwelstr. 30, D-5100 Aachen

Rizzardo, E., CSIRO Division of Chemicals and Polymers, Private Bag 10, Clayton, Vic, Australia 3168

Rozema, F.R., Dept. of Oral and Maxillofacial Surgery, University Hospital Groningen, Oostersingel 59, NL-9713 EZ Groningen

Schädlich, G., Helmholtz-Institute for Biomedical Engineering, Pauwelstr. 30, D-5100 Aachen

Scherer, M.A., Institut für Experimentelle Chirurgie der TU München, Ismaninger Str. 22, D-8000 München 80

Schindhelm, K., Centre for Biomedical Engineering, University of New South Wales, Kensington, NSW, Australia 2033

Schwengler, H., Beiersdorf AG, Unnastr. 48, D-2000 Hamburg 20

Shimizu, Y., Research Center for Biomedical Engineering, Kyoto University, 606 Kyoto, Sakyo-Ku, Japan

Siebels, W., Institut für Experimentelle Chirurgie der TU München, Ismaninger Str. 22, D-8000 München 80

Stokes, K.B., Medtronic Inc., 7000 Central Ave., Minneapolis, MN 55432, USA

Strahl, B., Beiersdorf AG, Unnastr. 48, D-2000 Hamburg 20

Thiede, A., Chirurgische Universitätsklinik Würzburg, Josef-Schneider-Str. 2, D-8700 Würzburg

Tyler, B.J., Dept. of Chemical Engineering and Center for Bioengineering, BF-10, University of Washington, Seattle, WA 98195, USA

Urbanski, P.W., Medtronic Inc., 7000 Central Ave., Minneapolis, MN 55432, USA

van der Elst, M., Free University Hospital, Department of Traumatology, P.O. Box 7057, NL-1007 MB Amsterdam

Wagner, B., Institut für Textil- und Verfahrenstechnik, Körschtalstr. 26, D-7306 Denkendorf

Welling, C., Beiersdorf AG, Unnastr. 48, D-2000 Hamburg 20

ACKNOWLEDGEMENT

The editors wish to thank the following companies for their financial support of the conference:

Ethicon GmbH, Norderstedt, FRG
Mercedes Benz AG, Stuttgart, FRG
E. Merck, Darmstadt, FRG
Vascutek Ltd., Inchinnan, GB

and the companies

Boehringer Ingelheim KG, Ingelheim, FRG
Carl Zeiss, Oberkochen, FRG

for the exhibition of their products.

MIX
Papier aus verantwortungsvollen Quellen
Paper from responsible sources
FSC® C105338

If you have any concerns about our products,
you can contact us on
ProductSafety@springernature.com

In case Publisher is established outside the EU,
the EU authorized representative is:
Springer Nature Customer Service Center GmbH
Europaplatz 3, 69115 Heidelberg, Germany

Printed by Libri Plureos GmbH
in Hamburg, Germany